Metal Foams: A Design Guide

Metal Foams: A Design Guide

M.F. Ashby, A.G. Evans, N.A. Fleck, L.J. Gibson, J.W. Hutchinson and H.N.G. Wadley

BOSTON OXFORD AUCKLAND JOHANNESBURG MELBOURNE NEW DELHI

A member of the Reed Elsevier group

Recognizing the importance of preserving what has been written, Butterworth–Heinemann prints its books on acid-free paper whenever possible.

Butterworth–Heinemann supports the efforts of American Forests and the Global ReLeaf program in its campaign for the betterment of trees, forests, and our environment.

ISBN 0-7506-7219-6

Library of Congress Cataloging-in-Publication Data
A complete record for this title is available from the Library of Congress

British Library Cataloguing-in-Publication Data
A complete record for this title is available from the British Library

The publisher offers special discounts on bulk orders of this book.
For information, please contact:
Manager of Special Sales
Butterworth–Heinemann
225 Wildwood Avenue
Woburn, MA 01801–2041
Tel: 781-904-2500
Fax: 781-904-2620
For information on all Butterworth–Heinemann publications available, contact our World Wide Web home page at: http://www.bh.com

10 9 8 7 6 5 4 3 2 1
Printed in the United States of America

Contents

Preface and acknowledgements

Metal foams are a new class of materials with low densities and novel physical, mechanical, thermal, electrical and acoustic properties. This Design Guide is a contribution to the concurrent development of their science and exploitation. It seeks to document design information for metal foams even as the scientific research and process development are evolving. This should help to identify promising industrial sectors for applications, guide process development and accelerate take-up.

This work is supported by the DARPA/ONR MURI Program through Grant No. N00014-1-96-1028 for Ultralight Metal Structures and by the British Engineering and Science Research Council through a Research Grant. Many individuals and groups have contributed to its contents. They include Professor B. Budiansky, Professor H. Stone, Professor R. Miller, Dr A. Bastawros, Dr Y. Sugimura of the Division of Engineering and Applied Sciences, Harvard University; Dr T.J. Lu, Dr Anne-Marie Harte, Dr V. Deshpande of the Micromechanics Centre, Engineering Department, Cambridge University; Dr E.W. Andrews and Dr L. Crews of the Department of Materials Science and Engineering, MIT; Professor D. Elzey, Dr D. Sypeck and Dr K. Dharmasena of the Department of Materials Science and Engineering, UVA; Dr John Banhart of the Fraunhofer Institüt Angewandte Materialsforschung, Bremen; Professor H.P. Degisher and Dr Brigdt Kriszt of the Technical University of Vienna, Dr Jeff Wood of Cymat Corp. Mississauga, Canada; and Mr Bryan Leyda of Energy Research and Generation Inc. Oakland, CA.

List of contributors

M.F. Ashby
Cambridge Centre for Micromechanics
Engineering Department
University of Cambridge
Cambridge CB2 1PZ
UK
mfa2@eng.cam.ac.uk

A.G. Evans
Princeton Materials Institute
Bowen Hall
70 Prospect Avenue
Princeton, NJ 08540
USA
anevans@princeton.edu

N.A. Fleck
Cambridge Centre for Micromechanics
Engineering Department
University of Cambridge
Cambridge CB2 1PZ
UK
naf1@eng.cam.ac.uk

L.J. Gibson
Department of Materials Science and Engineering
Massachusetts Institute of Technology
Cambridge, MA 02139
USA
ljgibson@mit.edu

J.W. Hutchinson
Division of Engineering and Applied Sciences
Harvard University

Oxford Street
Cambridge, MA 02138
USA
Hutchinson@mems.harvard.edu

H.N.G. Wadley
Department of Materials Science and Engineering
School of Engineering and Applied Science
University of Virginia
Charlottesville, VA 22903
USA
haydn@virginia.edu

Table of physical constants and conversion units

Physical constants in SI units

Absolute zero temperature	−273.2°C
Acceleration due to gravity, g	9.807 m/s^2
Avogadro's number, N_A	6.022×10^{23}
Base of natural logarithms, e	2.718
Boltzmann's constant, k	1.381×10^{-23} J/K
Faraday's constant, k	9.648×10^{4} C/mol
Gas constant, $\overline{R}$	8.314 J/mol/K
Permeability of vacuum, μ_0	1.257×10^{-6} H/m
Permittivity of vacuum, ε_0	8.854×10^{-12} F/m
Planck's constant, h	6.626×10^{-34} J/s
Velocity of light in vacuum, c	2.998×10^{8} m/s
Volume of perfect gas at STP	22.41×10^{-3} m^3/mol

Conversion of units

Angle, θ	1 rad	57.30°
Density, ρ	1 lb/ft^3	16.03 kg/m^3
Diffusion coefficient, D	1 cm^3/s	1.0×10^{-4} m^2/s
Force, F	1 kgf	9.807 N
	1 lbf	4.448 N
	1 dyne	1.0×10^{-5} N
Length, l	1 ft	304.8 mm
	1 inch	25.40 mm
	1 Å	0.1 nm
Mass, M	1 tonne	1000 kg
	1 short ton	908 kg
	1 long ton	1107 kg
	1 lb mass	0.454 kg
Specific heat, C_p	1 cal/g.°C	4.188 kJ/kg.°C
	Btu/lb.°F	4.187 kJ/kg.°C

Conversion of units

Stress intensity, K_{IC}	1 ksi$\sqrt{in}$	1.10 $MN/m^{3/2}$
Surface energy, γ	1 erg/cm^2	1 mJ/m^2
Temperature, T	1°F	0.556°K
Thermal conductivity, λ	1 cal/s.cm.°C	418.8 W/m.°C
	1 Btu/h.ft.°F	1.731 W/m.°C
Volume, V	1 Imperial gall	4.546×10^{-3} m^3
	1 US gall	3.785×10^{-3} m^3
Viscosity, η	1 poise	0.1 $N.s/m^2$
	1 lb ft.s	0.1517 $N.s/m^2$

Conversion of units – stress and pressure*

	MN/m^2	dyn/cm^2	lb/in^2	kgf/mm^2	bar	long ton/in^2
MN/m^2	1	10^7	1.45×10^2	0.102	10	6.48×10^{-2}
dyn/cm^2	10^{-7}	1	1.45×10^{-5}	1.02×10^{-8}	10^{-6}	6.48×10^{-9}
lb/in^2	6.89×10^{-3}	6.89×10^4	1	703×10^{-4}	6.89×10^{-2}	4.46×10^{-4}
kgf/mm^2	9.81	9.81×10^7	1.42×10^3	1	98.1	63.5×10^{-2}
bar	0.10	10^6	14.48	1.02×10^{-2}	1	6.48×10^{-3}
long ton/in^2	15.44	1.54×10^8	2.24×10^3	1.54	1.54×10^2	1

Conversion of units – energy*

	J	erg	cal	eV	Btu	ft lbf
J	1	10^7	0.239	6.24×10^{18}	9.48×10^{-4}	0.738
erg	10^{-7}	1	2.39×10^{-8}	6.24×10^{11}	9.48×10^{-11}	7.38×10^{-8}
cal	4.19	4.19×10^7	1	2.61×10^{19}	3.97×10^{-3}	3.09
eV	1.60×10^{-19}	1.60×10^{-12}	3.38×10^{-20}	1	1.52×10^{-22}	1.18×10^{-19}
Btu	1.06×10^3	1.06×10^{10}	2.52×10^2	6.59×10^{21}	1	7.78×10^2
ft lbf	1.36	1.36×10^7	0.324	8.46×10^{18}	1.29×10^{-3}	1

Conversion of units – power*

	kW (kJ/s)	erg/s	hp	ft lbf/s
kW (kJ/s)	1	10^{-10}	1.34	7.38×10^2
erg/s	10^{-10}	1	1.34×10^{-10}	7.38×10^{-8}
hp	7.46×10^{-1}	7.46×10^9	1	5.50×10^2
ft lbf/s	1.36×10^{-3}	1.36×10^7	1.82×10^{-3}	1

*To convert row unit to column unit, multiply by the number at the column–row intersection, thus $1 MN/m^2 = 10$ bar

Chapter 1

Introduction

Metal foams are a new, as yet imperfectly characterized, class of materials with low densities and novel physical, mechanical, thermal, electrical and acoustic properties. They offer potential for lightweight structures, for energy absorption, and for thermal management; and some of them, at least, are cheap. The current understanding of their production, properties and uses in assembled in this Design Guide. The presentation is deliberately kept as simple as possible. Section 1.1 expands on the philosophy behind the Guide. Section 1.2 lists potential applications for metal foams. Section 1.3 gives a short bibliography of general information sources; further relevant literature is given in the last section of each chapter.

At this point in time most commercially available metal foams are based on aluminum or nickel. Methods exist for foaming magnesium, lead, zinc, copper, bronze, titanium, steel and even gold, available on custom order. Given the intensity of research and process development, it is anticipated that the range of available foams will expand quickly over the next five years.

1.1 This Design Guide

Metallic foams ('metfoams') are a new class of material, unfamiliar to most engineers. They are made by a range of novel processing techniques, many still under development, which are documented in Chapter 2. At present metfoams are incompletely characterized, and the processes used to make them are imperfectly controled, resulting in some variability in properties. But even the present generation of metfoams have property profiles with alluring potential, and the control of processing is improving rapidly. Metfoams offer significant performance gains in light, stiff structures, for the efficient absorption of energy, for thermal management and perhaps for acoustic control and other, more specialized, applications (Section 1.2). They are recyclable and non-toxic. They hold particular promise for market penetration in applications in which several of these features are exploited simultaneously.

But promise, in today's competitive environment, is not enough. A survey of the history of development of new material suggests a scenario like that sketched in Figure 1.1. Once conceived, research on the new material accelerates rapidly, driven by scientific curiosity and by the often over-optimistic

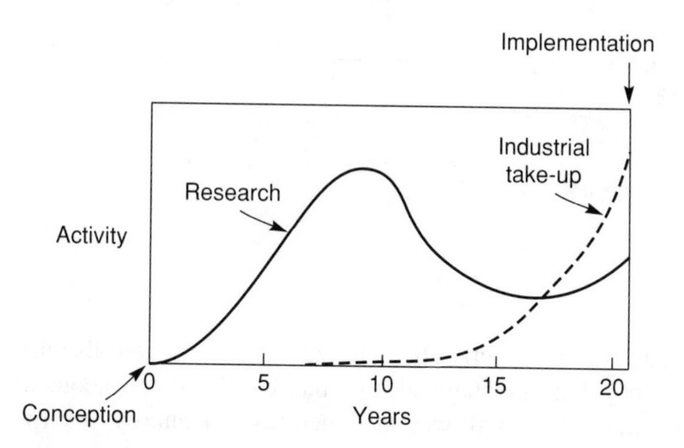

Figure 1.1 *A development history typical of many new materials. Research into the new material grows rapidly, and then slumps when little interest is shown by industry in using it. On a longer (15-year) time scale, applications slowly emerge*

predictions of its potential impact on engineering. The engineering take-up, however, is slow, held back by lack of adequate design data, experience and confidence; the disappointing take-up leads, after some 5 or 10 years, to disillusionment and a decline in research funding and activity. On a longer time-scale (15 years is often cited as the typical gestation period) the use of the new material – provided it has real potential – takes hold in one or more market sectors, and production and use expands, ultimately pulling research and development programmes with it.

There are obvious reasons for seeking a better balance between research and engineering take-up. This Design Guide is one contribution to the effort to achieve faster take-up, to give development curves more like those of Figure 1.2. Its seeks to do this by

- Presenting the properties of metallic foams in a way which facilitates comparison with other materials and structures
- Summarizing guidelines for design with them
- Illustrating how they might be used in lightweight structures, energy-absorbing systems, thermal management and other application, using, where possible, case studies.

The Guide starts with a description of the ways in which metfoams are made (Chapter 2) and the methods and precautions that have evolved for testing and characterizing them (Chapter 3). It continues with a summary of material properties, contrasting those of metfoams with those of other structural materials (Chapter 4). Chapter 5 outlines design analysis for materials selection. This is followed in Chapter 6 by a summary of formulae for simple structural shapes

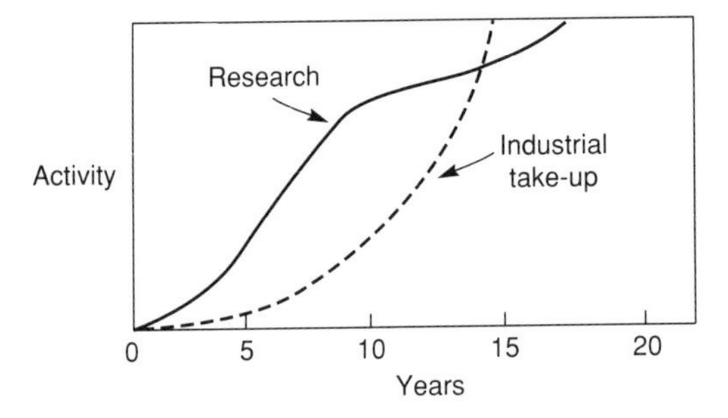

Figure 1.2 *A more attractive development history than that of Figure 1.1. Early formulation of design rules, research targeted at characterizing the most useful properties, and demonstrator projects pull the 'take-up' curve closer to the 'research' curve*

and loadings; the ways in which the properties of metal foams influence the use of these formulae are emphasized.

Mechanical design with foams requires constitutive equations defining the shape of the yield surface, and describing response to cyclic loading and to loading at elevated temperatures. These are discussed in Chapters 7, 8 and 9. One potential application for foams is that as the core for sandwich beams, panels and shells. Chapter 10 elaborates on this, illustrating how the stiffness and strength of weight-optimized sandwich structures compare with those of other types. Chapters 11, 12 and 13 outline the use of metal foams in energy, acoustic and thermal management. Chapter 14 describes how they can be cut, finished and joined. Chapter 15 discusses economic aspects of metal foams and the way economic and technical assessment are combined to establish viability. Chapter 16 reports case studies illustrating successful and potential applications of metal foams. Chapter 17 contains a list of the suppliers of metal foams, with contact information. Chapter 18 lists Web sites of relevant research groups and suppliers. The Guide ends with an Appendix in which material indices are catalogued.

1.2 Potential applications for metal foams

Application	Comment
Lightweight structures	Excellent stiffness-to-weight ratio when loaded in bending: attractive values of $E^{1/3}/\rho$ and $\sigma_y^{1/2}/\rho$ – see Chapter 5 and Appendix

Continued on next page

Application	Comment
Sandwich cores	Metal foams have low density with good shear and fracture strength – see Chapters 7 and 10
Strain isolation	Metal foams can take up strain mismatch by crushing at controled pressure – see Chapters 7 and 11
Mechanical damping	The damping capacity of metal foams is larger than that of solid metals by up to a factor of 10 – see Chapter 4
Vibration control	Foamed panels have higher natural flexural vibration frequencies than solid sheet of the same mass per unit area – see Chapter 4
Acoustic absorption	Reticulated metal foams have sound-absorbing capacity – see Chapter 12
Energy management: compact or light energy absorbers	Metal foams have exceptional ability to absorb energy at almost constant pressure – see Chapter 11
Packaging with high-temperature capability	Ability to absorb impact at constant load, coupled with thermal stability above room temperature – see Chapter 11
Artificial wood (furniture, wall panels)	Metal foams have some wood-like characteristics: light, stiff, and ability to be joined with wood screws – see Chapter 14
Thermal management: heat exchangers/ refrigerators	Open-cell foams have large accessible surface area and high cell-wall conduction giving exceptional heat transfer ability – see Chapter 13
Thermal management: flame arresters	High thermal conductivity of cell edges together with high surface area quenches combustion – see Chapter 13
Thermal management: heat shields	Metfoams are non-flammable; oxidation of cell faces of closed-cell aluminum foams appears to impart exceptional resistance to direct flame
Consumable cores for castings	Metfoams, injection-molded to complex shapes, are used as consumable cores for aluminum castings

Application	Comment
Biocompatible inserts	The cellular texture of biocompatible metal foams such as titanium stimulate cell growth
Filters	Open-cell foams with controled pore size have potential for high-temperature gas and fluid filtration
Electrical screening	Good electrical conduction, mechanical strength and low density make metfoams attractive for screening
Electrodes, and catalyst carriers	High surface/volume ratio allows compact electrodes with high reaction surface area – see Chapter 17
Buoyancy	Low density and good corrosion resistance suggests possible floatation applications

1.3 The literature on metal foams

The body of literature on metal foams is small, but growing quickly. The selection below gives sources that provide a general background. Specific references to more specialized papers and reports are given at the end of the chapter to which they are relevant.

Banhart, J. (1997) (ed.), *Metallschäume*, MIT Verlag, Bremen, Germany: the proceedings of a conference held in Bremen in March 1997, with extensive industrial participation (in German).

Banhart, J., Ashby, M.F. and Fleck, N.A. (eds), (1999) *Metal Foams and Foam Metal Structures*, Proc. Int. Conf. Metfoam'99, 14–16 June 1999, Bremen, Germany, MIT Verlag: the proceedings of a conference held in Bremen in June 1999 with extensive industrial participation (in English).

Evans, A.G. (ed.) (1998) *Ultralight Metal Structures*, Division of Applied Sciences, Harvard University, Cambridge, MA, USA: the annual report on the MURI programme sponsored by the Defence Advanced Research Projects Agency and Office of Naval Research.

Gibson, L.J. and Ashby, M.F. (1997) *Cellular Solids, Structure and Properties*, 2nd edition, Cambridge University Press, Cambridge, UK: a text dealing with mechanical, thermal, electrical and structural properties of foams of all types.

Shwartz, D.S., Shih, D.S., Evans, A.G. and Wadley, H.N.G. (eds) (1998) *Porous and Cellular Materials for Structural Application*, Materials Reseach Society Proceedings Vol. 521, MRS, Warrendale, PA, USA: the proceedings of a research conference containing a broad spectrum of papers on all aspects of metal foams.

Chapter 2
Making metal foams

Nine distinct process-routes have been developed to make metal foams, of which five are now established commercially. They fall into four broad classes: those in which the foam is formed from the vapor phase; those in which the foam is electrodeposited from an aqueous solution; those which depend on liquid-state processing; and those in which the foam is created in the solid state. Each method can be used with a small subset of metals to create a porous material with a limited range of relative densities and cell sizes. Some produce open-cell foams, others produce foams in which the majority of the cells are closed. The products differ greatly in quality and in price which, today, can vary from $7 to $12 000 per kg.

This chapter details the nine processes. Contact details for suppliers can be found in Chapter 17.

2.1 Making metal foams

The properties of metal foam and other cellular metal structures depend upon the properties of the metal, the relative density and cell topology (e.g. open or closed cell, cell size, etc.). Metal foams are made by one of nine processes, listed below. Metals which have been foamed by a given process (or a variant of it) are listed in square brackets.

1. Bubbling gas through molten Al–SiC or Al–Al_2O_3 alloys. [Al, Mg]
2. By stirring a foaming agent (typically TiH_2) into a molten alloy (typically an aluminum alloy) and controling the pressure while cooling. [Al]
3. Consolidation of a metal powder (aluminum alloys are the most common) with a particulate foaming agent (TiH_2 again) followed by heating into the mushy state when the foaming agent releases hydrogen, expanding the material. [Al, Zn, Fe, Pb, Au]
4. Manufacture of a ceramic mold from a wax or polymer-foam precursor, followed by burning-out of the precursor and pressure infiltration with a molten metal or metal powder slurry which is then sintered. [Al, Mg, Ni–Cr, stainless steel, Cu]
5. Vapor phase deposition or electrodeposition of metal onto a polymer foam precursor which is subsequently burned out, leaving cell edges with hollow cores. [Ni, Ti]

6. The trapping of high-pressure inert gas in pores by powder hot isostatic pressing (HIPing), followed by the expansion of the gas at elevated temperature. [Ti]
7. Sintering of hollow spheres, made by a modified atomization process, or from metal-oxide or hydride spheres followed by reduction or dehydridation, or by vapor-deposition of metal onto polymer spheres. [Ni, Co, Ni–Cr alloys]
8. Co-pressing of a metal powder with a leachable powder, or pressure-infiltration of a bed of leachable particles by a liquid metal, followed by leaching to leave a metal-foam skeleton. [Al, with salt as the leachable powder]
9. Dissolution of gas (typically, hydrogen) in a liquid metal under pressure, allowing it to be released in a controled way during subsequent solidification. [Cu, Ni, Al]

Only the first five of these are in commercial production. Each method can be used with a small subset of metals to create a porous material with a limited range of relative densities and cell sizes. Figure 2.1 summarizes the ranges of cell size, cell type (open or closed), and relative densities that can be manufactured with current methods.

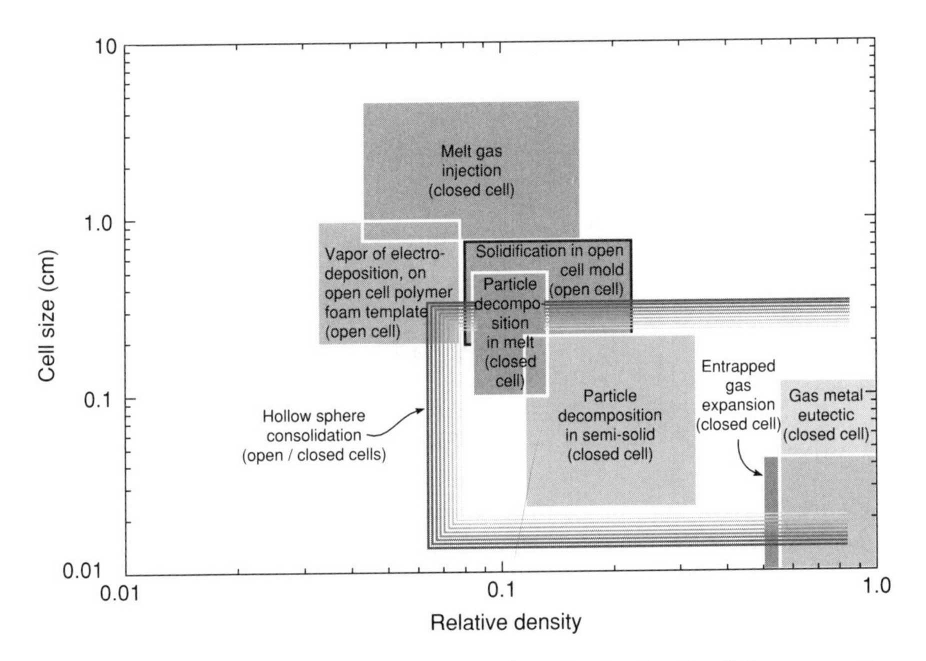

Figure 2.1 *The range of cell size and relative density for the different metal foam manufacturing methods*

2.2 Melt gas injection (air bubbling)

Pure liquid metals cannot easily be caused to foam by bubbling a gas into them. Drainage of liquid down the walls of the bubbles usually occurs too quickly to create a foam that remains stable long enough to solidify. However, 10–30% of small, insoluble, or slowly dissolving particles, such as aluminum oxide or silicon carbide, raise the viscosity of the aluminum melt and impede drainage in the bubble membrane, stabilizing the foam. Gas-injection processes are easiest to implement with aluminum alloys because they have a low density and do not excessively oxidize when the melt is exposed to air or other gases containing oxygen. There are several variants of the method, one of which is shown in Figure 2.2. Pure aluminum or an aluminum alloy is melted and 5–15 wt% of the stabilizing ceramic particles are added. These particles, typically 0.5–25 μm in diameter, can be made of alumina, zirconia, silicon carbide, or titanium diboride.

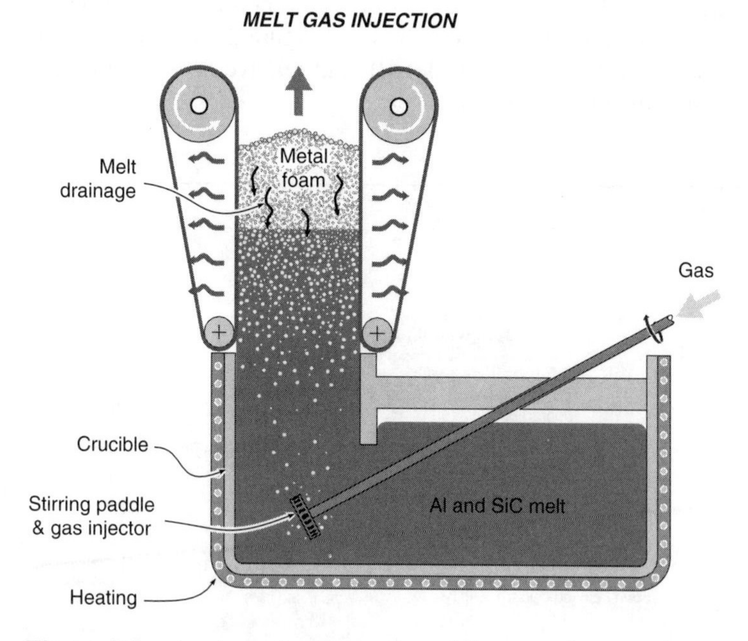

Figure 2.2 *A schematic illustration of the manufacture of an aluminum foam by the melt gas injection method (CYMAT and HYDRO processes)*

A variety of gases can be used to create bubbles within liquid aluminum. Air is most commonly used but carbon dioxide, oxygen, inert gases, and even water can also be injected into liquid aluminum to create bubbles. Bubbles formed by this process float to the melt surface, drain, and then begin to solidify. The thermal gradient in the foam determines how long the foam remains liquid or semi-solid, and thus the extent of drainage. Low relative density, closed-cell foams can be produced by carefully controling the gas-injection process and the cooling rate of the foam.

Various techniques can be used to draw-off the foam and create large (up to 1 m wide and 0.2 m thick slabs containing closed cell pores with diameters between 5 and 20 mm. NORSK-HYDRO and CYMAT (the latter using a process developed by ALCAN in Canada) supply foamed aluminum alloys made this way. This approach is the least costly to implement and results in a foam with relative densities in the range 0.03 to 0.1. It is at present limited to the manufacture of aluminum foams.

2.3 Gas-releasing particle decomposition in the melt

Metal alloys can be foamed by mixing into them a foaming agent that releases gas when heated. The widely used foaming agent titanium hydride (TiH_2) begins to decompose into Ti and gaseous H_2 when heated above about 465°C. By adding titanium hydride particles to an aluminum melt, large volumes of hydrogen gas are rapidly produced, creating bubbles that can lead to a closed-cell foam, provided foam drainage is sufficiently slow, which requires a high melt viscosity. The Shinko Wire Company has developed an aluminum foam trade named Alporas using this approach (Figure 2.3).

The process begins by melting aluminum and stabilizing the melt temperature between 670 and 690°C. Its viscosity is then raised by adding 1–2% of calcium which rapidly oxidizes and forms finely dispersed CaO and $CaAl_2O_4$ particles. The melt is then aggressively stirred and 1–2% of TiH_2 is added in the form of 5–20 μm diameter particles. As soon as these are dispersed in the melt, the stirring system is withdrawn, and a foam is allowed to form above the melt. Control of the process is achieved by adjusting the overpressure, temperature and time. It takes, typically, about ten minutes to totally decompose the titanium hydride. When foaming is complete the melt is cooled to solidify the foam before the hydrogen escapes and the bubbles coalesce or collapse.

The volume fraction of calcium and titanium hydride added to the melt ultimately determines the relative density and, in combination with cooling conditions, the cell size. The cell size can be varied from 0.5 to 5 mm by changing the TiH_2 content, and the foaming and cooling conditions. Relative densities from 0.2 to as low as 0.07 can be manufactured. As produced, the Alporas foam has predominantly closed cells, though a subsequent rolling treatment can be used to fracture many of the cell walls in order to increase their acoustic damping. A significant manufacturing capacity now exists in Japan. Although only small volume fractions of expensive calcium and titanium hydride are used, the process is likely to be more costly than gas-injection

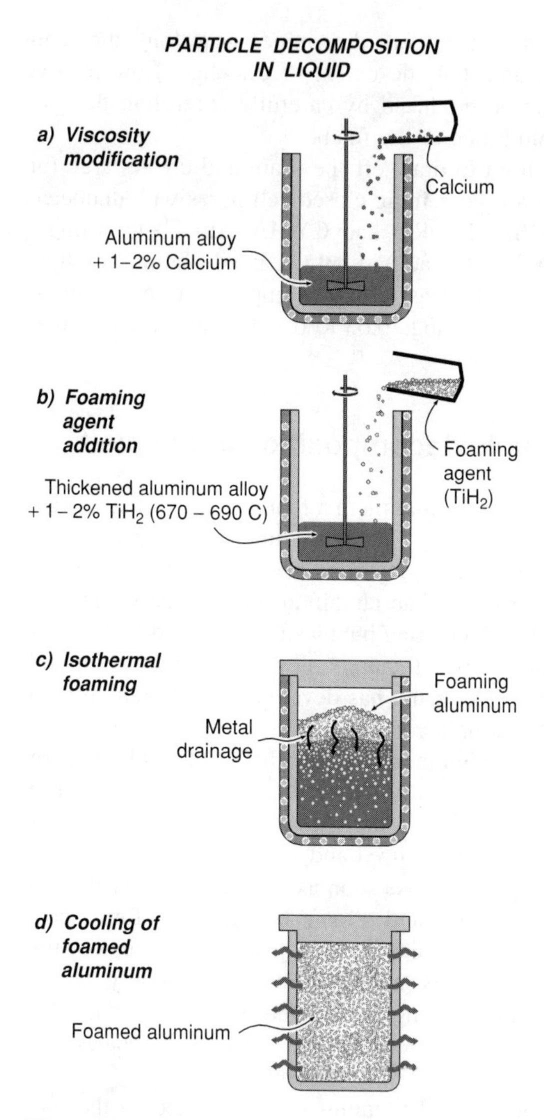

Figure 2.3 *The process steps used in the manufacture of aluminum foams by gas-releasing particle decomposition in the melt (Alporas process)*

methods because it is a batch process. Today, only aluminum alloys are made in this way because hydrogen embrittles many metals and because the decomposition of TiH_2 occurs too quickly in higher melting point alloys. Research using alternative foaming agents (carbonates, nitrates) with higher decomposition temperatures offers the prospect of using this method to foam iron, steels and nickel-based alloys.

2.4 Gas-releasing particle decomposition in semi-solids

Foaming agents can be introduced into metals in the solid state by mixing and consolidating powders. Titanium hydride, a widely used foaming agent, begins to decompose at about 465°C, which is well below the melting point of pure aluminum (660°C) and of its alloys. This raises the possibility of creating a foam by dispersing the foaming agent in solid aluminum using powder metallurgy processes and then raising the temperature sufficiently to cause gas release and partial or full melting of the metal, allowing bubble growth. Cooling then stabilizes the foam. Several groups, notably IFAM in Bremen, Germany, LKR in Randshofen, Austria, and Neuman-Alu in Marktl, Austria, have developed this approach.

A schematic diagram of the manufacturing sequence is shown in Figure 2.4. It begins by combining particles of a foaming agent (typically titanium hydride) with an aluminum alloy powder. After the ingredients are thoroughly mixed, the powder is cold compacted and then extruded into a bar or plate of near theoretical density. This 'precursor' material is chopped into small pieces, placed inside a sealed split mold, and heated to a little above the solidus temperature of the alloy. The titanium hydride then decomposes, creating voids with a high internal pressure. These expand by semi-solid flow and the aluminum swells, creating a foam that fills the mold. The process results in components with the same shape as the container and relative densities as low as 0.08. The foam has closed cells with diameters that range from 1 to 5 mm in diameter.

IFAM, Bremen, have developed a variant of the process, which has considerable potential for innovative structural use. Panel structures are made by first roll-bonding aluminum or steel face-sheets onto a core-sheet of unexpanded precursor. The unexpanded sandwich structure is then pressed or deep-drawn to shape and placed in a furnace to expand the core, giving a shaped, metal-foam cored sandwich-panel. Only foamed aluminum is commercially available today, but other alloy foams are being developed using different foaming agents.

2.5 Casting using a polymer or wax precursor as template

Open-cell polymer foams with low relative densities and a wide range of cell sizes of great uniformity are available from numerous sources. They can be used as templates to create investment-casting molds into which a variety of metals and their alloys can be cast. It is thought that the ERG DUOCEL range of foams are made in this way. The method is schematically illustrated in Figure 2.5.

PARTICLE DECOMPOSITION IN SEMI-SOLID

a) Select Ingredients

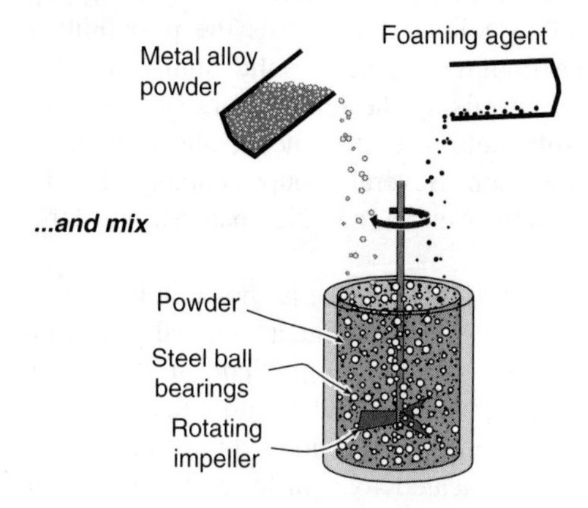

b) Consolidation & Extrusion

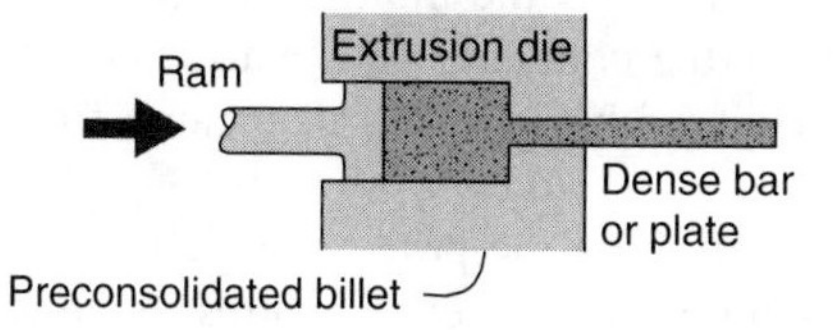

c) Shaped mold

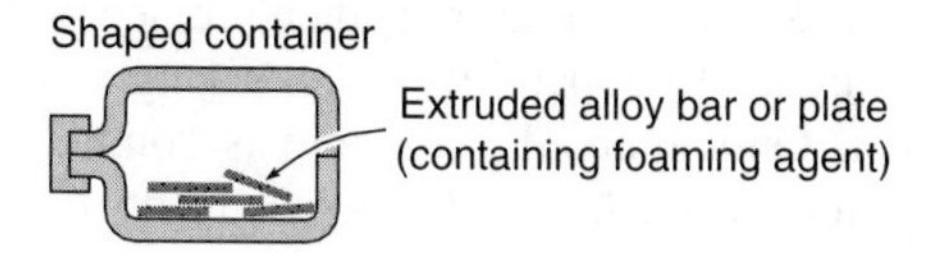

d) Foaming

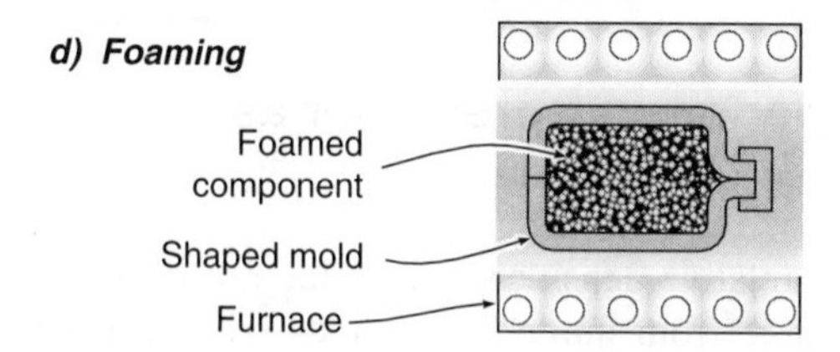

Figure 2.4 *The sequence of powder metallurgy steps used to manufacture metal foams by gas-releasing particles in semi-solids (the Fraunhofer and the Alulight processes)*

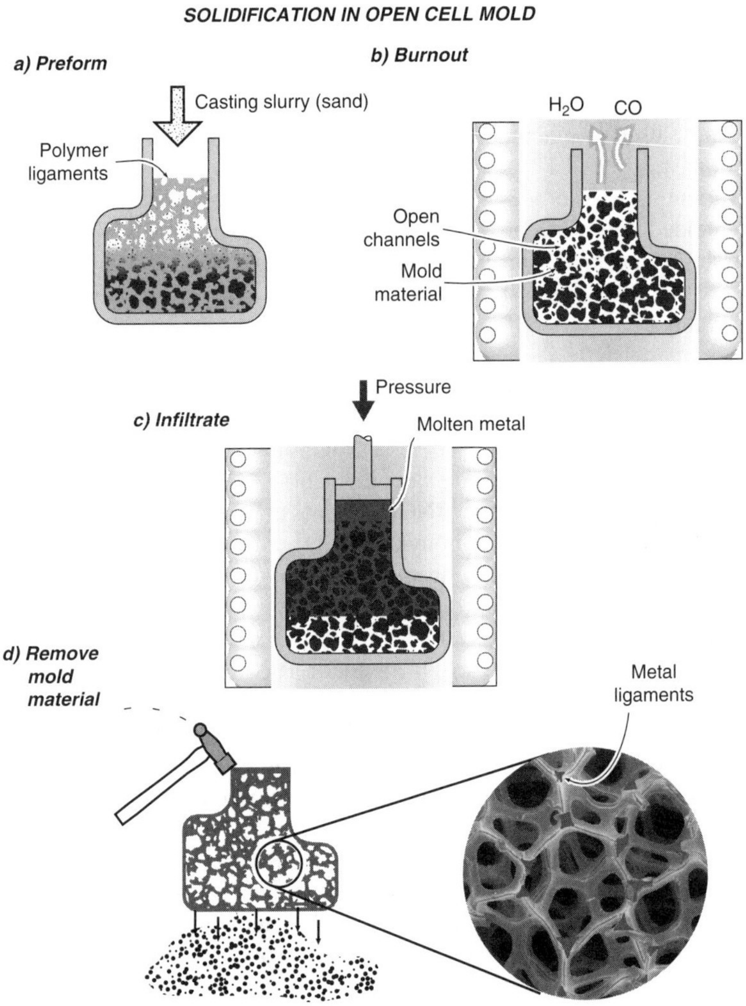

SOLIDIFICATION IN OPEN CELL MOLD

a) Preform

Casting slurry (sand)

Polymer ligaments

b) Burnout

H_2O CO

Open channels

Mold material

Pressure

c) Infiltrate

Molten metal

d) Remove mold material

Metal ligaments

Figure 2.5 *Investment casting method used to manufacture open cell foams (DUOCEL process)*

An open-cell polymer foam mold template with the desired cell size and relative density is first selected. This can be coated with a mold casting (ceramic powder) slurry which is then dried and embedded in casting sand. The mold is then baked both to harden the casting material and to decompose (and evaporate) the polymer template, leaving behind a negative image of the foam. This mold is subsequently filled with a metal alloy and allowed to cool. The use of a moderate pressure during melt infiltration can overcome the resistance to flow of some liquid alloys. After directional solidification and cooling, the mold materials are removed leaving behind the metal equivalent of the original polymer foam. Metal powder slurries can also be used instead of liquid metals. These are subsequently sintered. The method gives open-cell foams with pore sizes of 1–5 mm and relative densities as low as 0.05. The process can be used to manufacture foams from almost any metal that can be investment cast.

In a variant of the process, the precursor structure is assembled from injection-molded polymeric or wax lattices. The lattice structure is coated with a casting slurry and fired, burning it out and leaving a negative image mold. Metal is cast or pressure-cast into the mold using conventional investment casting techniques.

2.6 Metal deposition on cellular preforms

Open-cell polymer foams can serve as templates upon which metals are deposited by chemical vapor decomposition (CVD), by evaporation or by electrodeposition. In the INCO process, nickel is deposited by the decomposition of nickel carbonyl, $Ni(CO)_4$. Figure 2.6 schematically illustrates one approach in which an open-cell polymer is placed in a CVD reactor and nickel carbonyl is introduced. This gas decomposes to nickel and carbon monoxide at a temperature of about 100°C and coats all the exposed heated surfaces within the reactor. Infrared or RF heating can be used to heat only the polymer foam. After several tens of micrometers of the metal have been deposited, the metal-coated polymer foam is removed from the CVD reactor and the polymer is burnt out by heating in air. This results in a cellular metal structure with hollow ligaments. A subsequent sintering step is used to densify the ligaments.

Nickel carbonyl gas is highly toxic and requires costly environmental controls before it can be used for manufacturing nickel foams. Some countries, such as the United States, have effectively banned its use and others make it prohibitively expensive to implement industrial processes that utilize nickel carbonyl gas. Electro- or electroless deposition methods have also been used to coat the preforms, but the nickel deposited by the CVD technique has a lower electrical resistance than that created by other methods. The pore size can be varied over a wide range. Foams with open pore sizes in the 100–300 μm diameter range are available. The method is restricted to pure elements such as nickel or titanium because of the difficulty of CVD or electrodeposition of alloys. It gives the lowest relative density (0.02–0.05) foams available today.

2.7 Entrapped gas expansion

The solubility in metals of inert gases like argon is very low. Powder metallurgy techniques have been developed to manufacture materials with a dispersion of small pores containing an inert gas at a high pressure. When these

METAL DEPOSITION
ON CELLULAR PREFORM

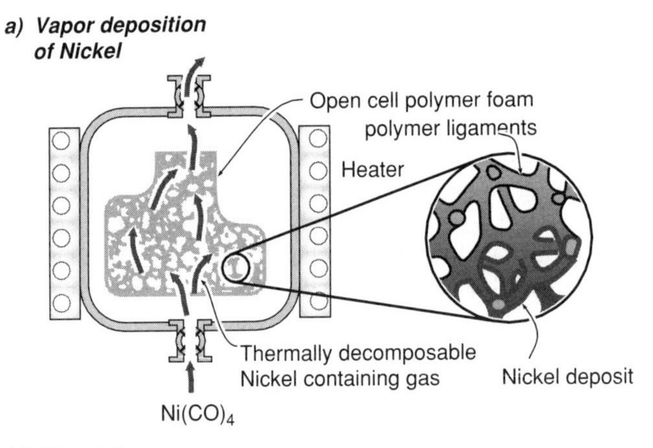

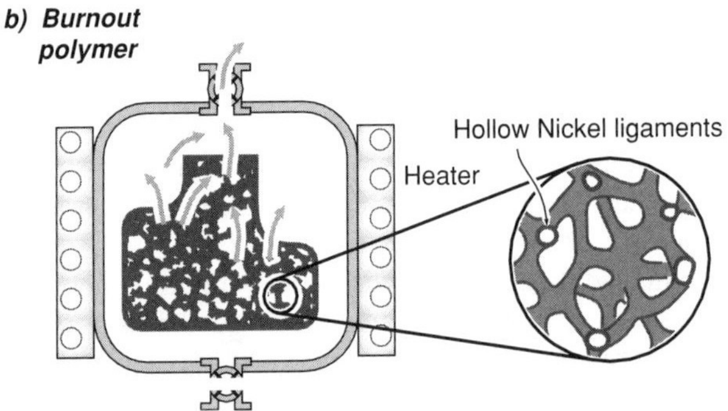

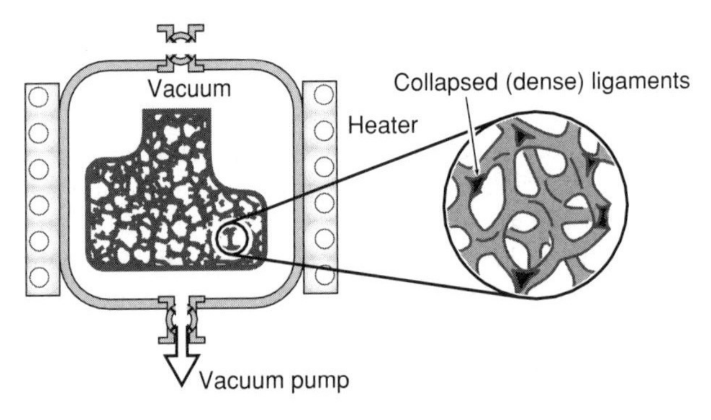

Figure 2.6 *Schematic illustration of the CVD process used to create open-cell nickel foams (INCO process)*

materials are subsequently heated, the pore pressure increases and the pores expand by creep of the surrounding metal (Figure 2.7). This process has been used by Boeing to create low-density core (LDC) Ti–6A1–4V sandwich panels with pore fractions up to 50%.

In the process Ti–6A1–4V powder is sealed in a canister of the same alloy. The canister is evacuated to remove any oxygen (which embrittles titanium) and then backfilled with between 3 to 5 atmospheres (0.3–0.5 MPa) of argon. The canister is then sealed and consolidated to a high relative density (0.9–0.98) by HIPing causing an eight-fold increase in void pressure. This is too low to cause expansion of Ti–6AI–4V at room temperature. The number of pores present in the consolidated sample is relatively low (it is comparable to the number of powder particles in the original compact), so a rolling step is introduced to refine the structure and create a more uniform distribution of small pores. In titanium alloys, rolling at 900–940°C results in void flattening and elongation in the rolling direction. As the voids flatten, void faces come into contact and diffusion bond, creating strings of smaller gas-filled pores. Cross-rolling improves the uniformity of their distribution. Various cold sheet forming processes can then be used to shape the as-rolled plates.

The final step in the process sequence is expansion by heating at 900°C for 20–30 hours. The high temperature raises the internal pore pressure by the ratio of the absolute temperature of the furnace to that of the ambient (about a factor of four) i.e. to between 10 and 16 MPa, causing creep dilation and a reduction in the overall density of the sample.

This process results in shaped Ti-alloy sandwich construction components with a core containing a closed-cell void fraction of up to 0.5 and a void size of 10–300 μm. While it shares most of the same process steps as P/M manufacturing, and the cost of the inert gas is minor, HIPing and multipass hot cross-rolling of titanium can be expensive. This process is therefore likely to result in materials that are more costly to manufacture than P/M alloys.

2.8 Hollow sphere structures

Several approaches have recently emerged for synthesizing hollow metal spheres. One exploits the observation that inert gas atomization often results in a small fraction (1–5%) of large-diameter (0.3–1 mm) hollow metal alloy spheres with relative densities as low as 0.1. These hollow particles can then be sorted by flotation methods, and consolidated by HIPing, by vacuum sintering, or by liquid-phase sintering. Liquid-phase sintering may be the preferred approach for some alloys since it avoids the compressive distortions of the thin-walled hollow powder particles that results from the HIPing process and avoids the prolonged high-temperature treatments required to achieve strong particle–particle bonds by vacuum sintering methods. Porous nickel

ENTRAPPED GAS EXPANSION

Process Steps

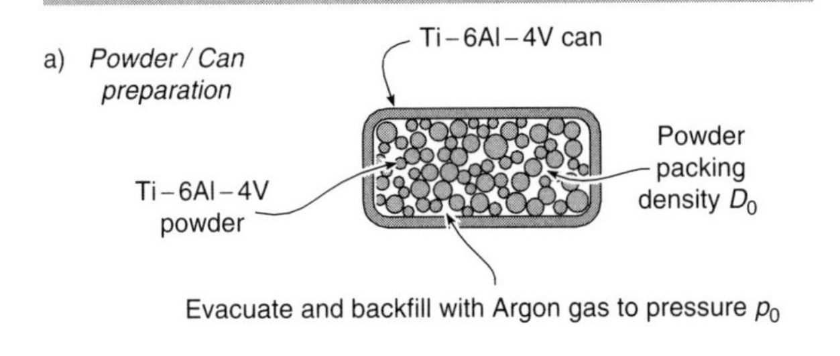

b) *HIP Consolidation*
(900°C, 100 – 200 MPa, 2 hrs.)

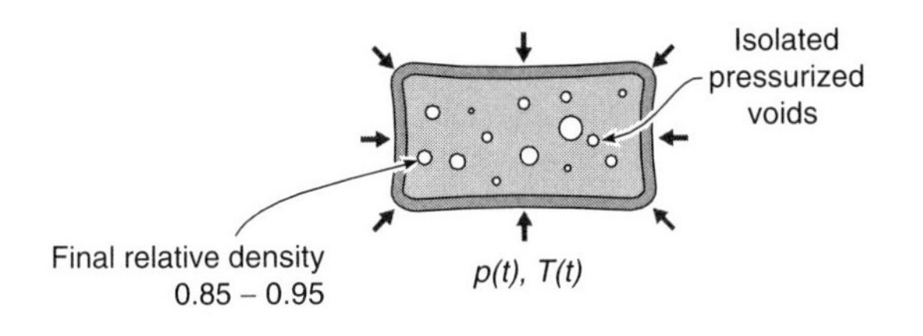

c) *Hot rolling*
(approx. 930°C, 6 – 40 passes in air)

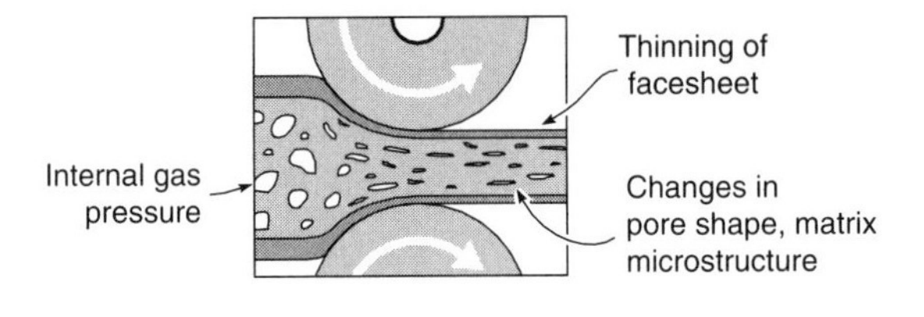

d) *Expansion heat treatment*
(900°C, 4 – 48 hrs.)

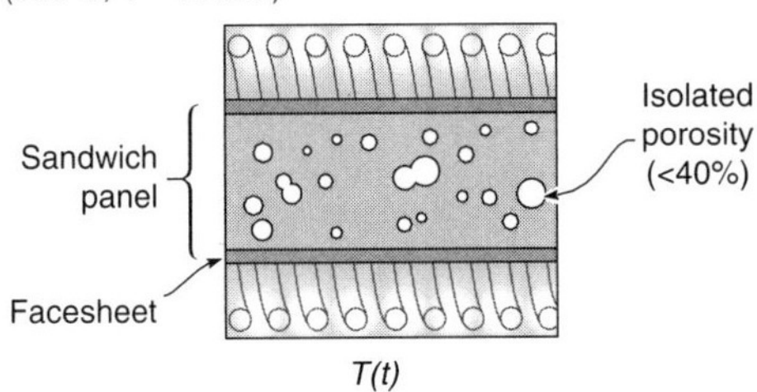

Figure 2.7 *Process steps used to manufacture titanium alloy sandwich panels with highly porous closed-cell cores*

superalloys and Ti–6Al–4V with relative densities of 0.06 can be produced in the laboratory using this approach. The development of controled hollow powder atomization techniques may enable economical fabrication of low-density alloy structures via this route.

In an alternative method, hollow spheres are formed from a slurry composed of a decomposable precursor such as TiH_2, together with organic binders and solvents (Figure 2.8). The spheres are hardened by evaporation during their

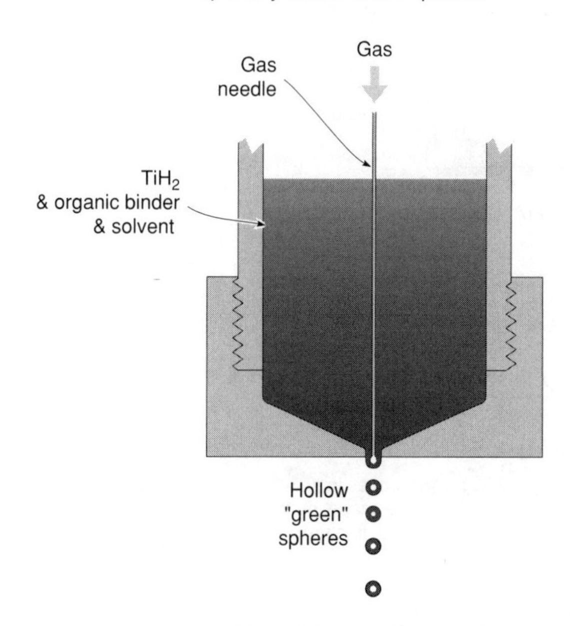

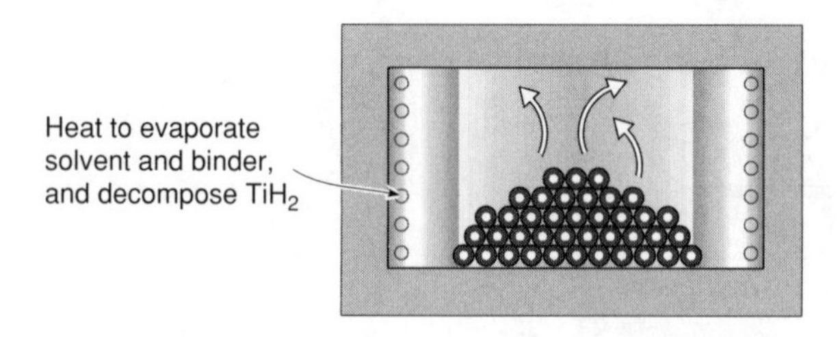

Figure 2.8 *The Georgia Tech route for creating hollow metal spheres and their consolidation to create a foam with open- and closed-cell porosity*

flight in a tall drop tower, heated to drive off the solvents and to volatilize the binder. A final heat treatment decomposes the metal hydride leaving hollow metal spheres. The approach, developed at Georgia Tech, can be applied to many materials, and is not limited to hydrides. As an example, an oxide mixture such as Fe_2O_3 plus Cr_2O_3 can be reduced to create a stainless steel.

In a third method developed at IFAM, Bremen, polystyrene spheres are coated with a metal slurry and sintered, giving hollow metal spheres of high uniformity. The consolidation of hollow spheres gives a structure with a mixture of open and closed porosity. The ratio of the two types of porosity and the overall relative density can be tailored by varying the starting relative density of the hollow spheres and the extent of densification during consolidation. Overall relative densities as low as 0.05 are feasible with a pore size in the range 100 μm to several millimetres.

2.9 Co-compaction or casting of two materials, one leachable

Two powders, neither with a volume fraction below 25%, are mixed and compacted, forming double-connected structures of both phases. After consolidation one powder (e.g. salt) is leached out in a suitable solvent (Figure 2.9). Foams based on powder mixes of aluminum alloys with sodium chloride have successfully been made in large sections with uniform structures. The resulting cell shapes differ markedly from those of foams made by other methods. In practice the method is limited to producing materials with relative densities

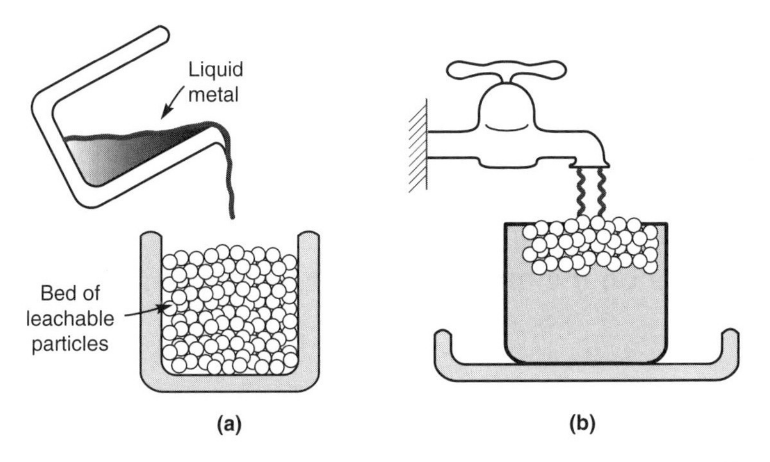

Figure 2.9 *(a) A bed of leachable particles (such as salt) is infiltrated with a liquid metal (such as aluminum or one of its alloys). (b) The particles are disolved in a suitable solvent (such as water) leaving an open-cell foam*

between 0.3 and 0.5. The cell size is determined by the powder particle size, and lies in the range 10 μm to 10 mm.

In an alternative but closely related process, a bed of particles of the leachable material is infiltrated by liquid metal under pressure, and allowed to cool. Leaching of the particles again gives a cellular metallic structure of great uniformity.

2.10 Gas–metal eutectic solidification

Numerous metal alloy–hydrogen binary phase diagrams exhibit a eutectic; these include Al-, Be-, Cr-, Cu-, Fe-, Mg-, Mn- and Ni-based alloys. The alloys are melted, saturated with hydrogen under pressure, and then directionally solidified, progressively reducing the pressure. During solidification, solid metal and hydrogen simultaneously form by a gas eutectic reaction, resulting in a porous material containing hydrogen-filled pores. These materials are referred to as GASARs (or GASERITE).

A schematic diagram of the basic approach is shown in Figure 2.10. A furnace placed within a pressure vessel is used to melt an alloy under an appropriate pressure of hydrogen (typically 5–10 atmospheres of hydrogen). This melt is then poured into a mold where directional eutectic solidification is allowed to occur. This results in an object containing a reasonably large (up to 30%) volume fraction of pores. The pore volume fraction and pore orientation are a sensitive function of alloy chemistry, melt over-pressure, melt superheat (which affects the hydrogen solubility of the liquid metal), the temperature field in the liquid during solidification, and the rate of solidification. With so many process variables, control and optimization of the pore structure are difficult. The method poses certain safety issues, and in its present form is a batch process. As a result, materials manufactured by this route are costly. Though GASAR materials were among the first highly porous materials to attract significant interest, they remain confined to the laboratory and are not yet commercially available.

2.11 Literature on the manufacture of metal foams

General

Astro Met, Inc. *Ampormat Porous Materials*, Astro Met, Inc., Cinncinnati.

Davies, G.J. and Zhen, S. (1983) Metallic foams: their production, properties, and applications. *Journal of Materials Science* **18** 1899–1911.

Banhart, J. and Baumeister, J. (1988) *Production methods for metallic foams.* In Shwartz, D.S., Shih, D.S., Evans, A.G. and Wadley, H.N.G. (eds) (1998) *Porous and Cellular Materials for Structural Application*, Materials Research Society Proceedings, Vol. 521, MRS, Warrendale, PA, USA.

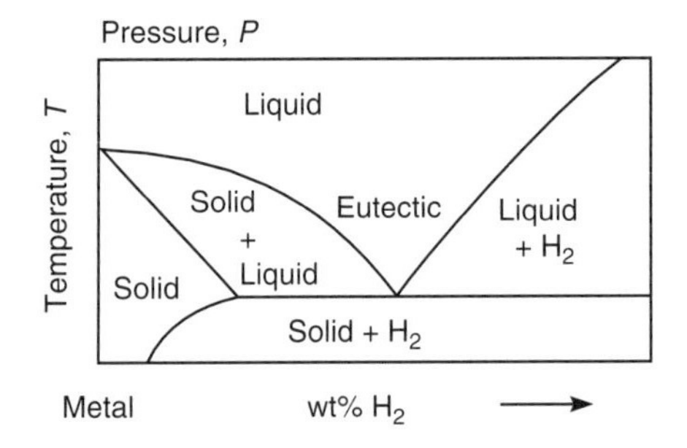

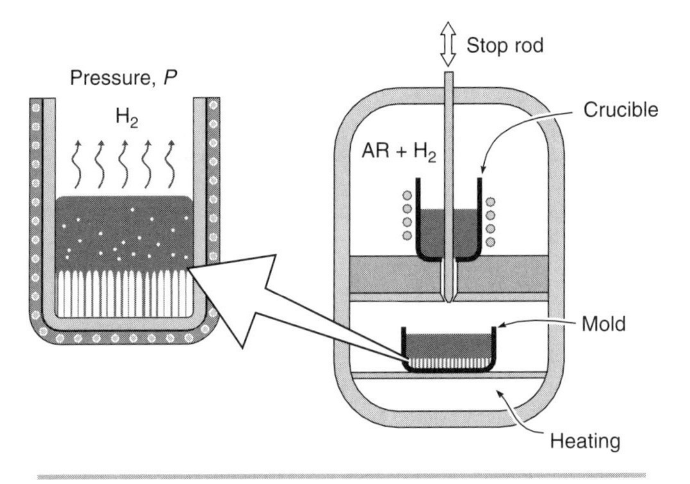

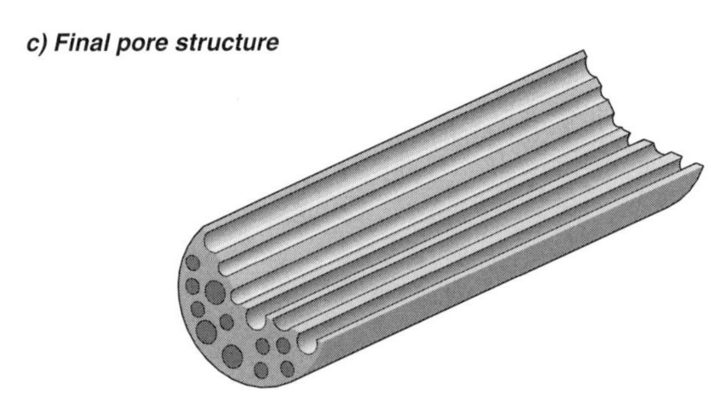

Figure 2.10 *Gas–metal eutectic solidification for the manufacture of GASARs*

Melt gas injection

Jin *et al.* (1990) Method of producing lightweight foamed metal. US Patent No. 4,973,358.

Jin *et al.* (1992) Stabilized metal foam body. US Patent No. 5, 112, 697.

Jin *et al.* (1993) Lightweight metal with isolated pores and its production. US Patent No. 5,221,324.

Kenny *et al.* (1994) Process for shape casting of particle stabilized metal foam. US Patent No. 5,281,251.

Niebyski *et al.* (1974) Preparation of metal foams with viscosity increasing gases. US Patent No. 3,816,952.

Sang *et al.* (1994) Process for producing shaped slabs of particle stabilized foamed metal. US Patent No. 5,334,236.

Thomas *et al.* (1997) Particle-stablilized metal foam and its production. US Patent No. 5,622,542.

Particle decomposition in melts

Akiyama *et al.* (1987) Foamed metal and method of producing same. US Patent No. 4,713,277.

Elliot, J.C. (1956) Method of producing metal foam. US Patent No. 2,751,289.

Gergely, V. and Clyne, T.W. (1998) The effect of oxide layers on gas-generating hydride particles during production of aluminum foams. In Shwartz, D.S., Shih, D.S., Evans, A.G. and Wadley, H.N.G. (eds) *Porous and Cellular Materials for Structural Application*, Materials Research Society Proceedings, Vol. 521, MRS, Warrendale, PA, USA.

Miyoshi, T., Itoh, M., Akiyama, S. and Kitahara, A. (1998) *Aluminum foam, ALPORAS, the production process, properties and applications*, Shinko Wire Company, Ltd, New Tech Prod. Div., Amagasaki, Japan.

Speed, S.F. (1976) Foaming of metal by the catalyzed and controled decomposition of zirconium hydride and titanium hydride. US Patent No. 3,981,720.

Particle decomposition in semisolids

Baumeister, J. (1988) Methods for manufacturing foamable metal bodies. US Patent 5,151,246.

Falahati, A. (1997) *Machbarkeitsstudie zur Herstellung von Eisenbasisschaum*, Master's thesis, Technical University, Vienna.

Yu, C.-J. and Eifert, H. (1998) Metal foams. *Advanced Materials & Processes* November, 45–47.

MEPURA. (1995) Alulight. Metallpulver GmbH. Brannau-Ranshofen, Austria.

Solidification in open-cell molds

ERG (1998) Duocel aluminum foam. *ERG Corporate Literature and Reports*, 29 September. <http://ergaerospace.com/lit.html>

ERG (1998) Duocel physical properties. *ERG Corporate Literature and Reports*, 29 September. <http://ergaerospace.com/lit.html>

H_2/metal eutectic solidification

Sharpalov, Yu (1993) Method for manufacturing porous articles. US Patent No. 5,181,549.

Sharpalov. (1994) Porous metals. *MRS Bulletin*, April: 24–28.

Zheng, Y., Sridhar, S. and Russell, K.C. (1996) Controled porosity alloys through solidification processing: a modeling study. *Mat. Res. Soc. Symp. Proc.* **371**: 365–398.

Vapor (electro) deposition on cellular preforms

Babjak *et al.* (1990) Method of forming nickel foam. US Patent No. 4,957,543.

Entrapped gas expansion

Elzey, D.M. and Wadley, H.N.G. (1998) The influence of internal pore pressure during roll forming of structurally porous metals. In *Porous and Cellular Materials for Structural Applications*, MRS, Warrendale, PA, USA.

Kearns, M.W., Blekinsop, P.A., Barber, A.C. and Farthing, T.W. (1988) Manufacture of a novel porous metal. *The International Journal of Powder Metallurgy* **24**: 59–64.

Kearns, M.W., Blekinsop, P.A., Barber, A.C. and Farthing, T.W. (1988) Novel porous titanium. Paper presented at Sixth World Conference on Titanium, Cannes, France.

Martin, R.L. and Lederich, R.J. (1991) Porous core/BE TI 6-4 development for aerospace structures. In *Advances in Powder Metallurgy: Proceeding of the 1991 Powder Metallurgy Conference and Exposition*, Powder Metallurgy Industries Federation, Princeton, NJ, USA.

Schwartz, D.S. and Shih, D.S. (1998) Titanium foams made by gas entrappment. In Shwartz, D.S., Shih, D.S., Evans, A.G. and Wadley, H.N.G. (eds) *Porous and Cellular Materials for Structural Application*, Materials Research Society Proceedings, Vol. 521, MRS, Warrendale PA, USA.

Hollow sphere consolidation

Drury, W.J., Rickles, S.A. Sanders, T.H. and Cochran, J.K. (1989) Deformation energy absorption characteristics of a metal/ceramic cellular solid. In *Proceedings of TMS Conference on Light Weight Alloys for Aerospace Applications*, TMS-AIME, Warrendale, PA, USA.

Kendall, J.M., Lee, M.C. and Wang, T.A. (1982) Metal shell technology based upon hollow jet instability. *Journal of Vacuum Science Technology* **20**, 1091–1093.

Lee *et al.* (1991) Method and apparatus for producing microshells. US Patent No. 5,055,240.

Sypeck, D.S., Parrish, P.A. and Wadley, H.N.G. (1998) Novel hollow powder porous structures. In *Porous and Cellular Materials for Structural Applications*, Materials Research Society Proceeding, Vol. 521, MRS, Warrendale, PA, USA.

Torobin (1983) Method and apparatus for producing hollow metal microspheres and microspheroids. US Patent No. 4,415,512.

Uslu, C., Lee, K.J. Sanders, T.H. and Cochran, J.K. (1997) Ti–6AI–4V hollow sphere foams. In *Synthesis/Processing of Light Weight Metallic Materials II*, TMS, Warrendale, PA, USA.

Wang *et al.* (1984) Apparatus for forming a continuous lightweight multicell material. US Patent No. 4,449,901.

Wang *et al.* (1982) Method and apparatus for producing gas-filled hollow spheres. US Patent No. 4,344,787.

Co-compaction or casting of two materials, one leachable

De Ping, H.E., Department of Materials Science and Engineering, Southeast University, No. 2, Sipailu, Nanjing 210096, PR China (dphe@seu.edu.cn)

Chapter 3

Characterization methods

The cellular structure of metallic foams requires that special precautions must be taken in characterization and testing. Structure is examined by optical microscopy, scanning electron microscopy and X-ray tomography. The apparent moduli and strength of foam test samples depends on the ratio of the specimen size to the cell size, and can be influenced by the state of the surface and the way in which the specimen is gripped and loaded. This means that specimens must be large (at least seven cell diameters of every dimension) and that surface preparation is necessary. Local plasticity at stresses well below the general yield of the foam requires that moduli be measured from the slope of the unloading curve, rather than the loading curve. In this chapter we summarize reliable methods for characterizing metallic foams in uniaxial compression, uniaxial tension, shear and multiaxial stress states, under conditions of creep and fatigue, and during indentation. An optical technique for measuring the surface displacement field, from which strains can be calculated, is described.

3.1 Structural characterization

A metal foam is characterized structurally by its *cell topology* (open cells, closed cells), *relative density*, *cell size* and *cell shape* and *anisotropy*. Density is best measured by weighing a sample of known volume; the rest require microscopy.

Optical microscopy is helpful in characterizing metal foams provided that the foam is fully impregnated with opaque epoxy (or equivalent) before polishing. This requires that the foam sample be immersed in a low-viscosity thermoset containing a coloring agent (black or deep blue is best), placed in a vacuum chamber and degassed and then repressurized to force the polymer into the cells. The procedure may have to be repeated for closed-cell foams after coarse polishing, since this often opens a previously closed cell. Conventional polishing then gives reliable sections for optical microscopy (Figure 3.1(a)).

Scanning electron microscopy (SEM) is straightforward; the only necessary precaution is that relating to surface preparation (see Section 3.2). SEM is

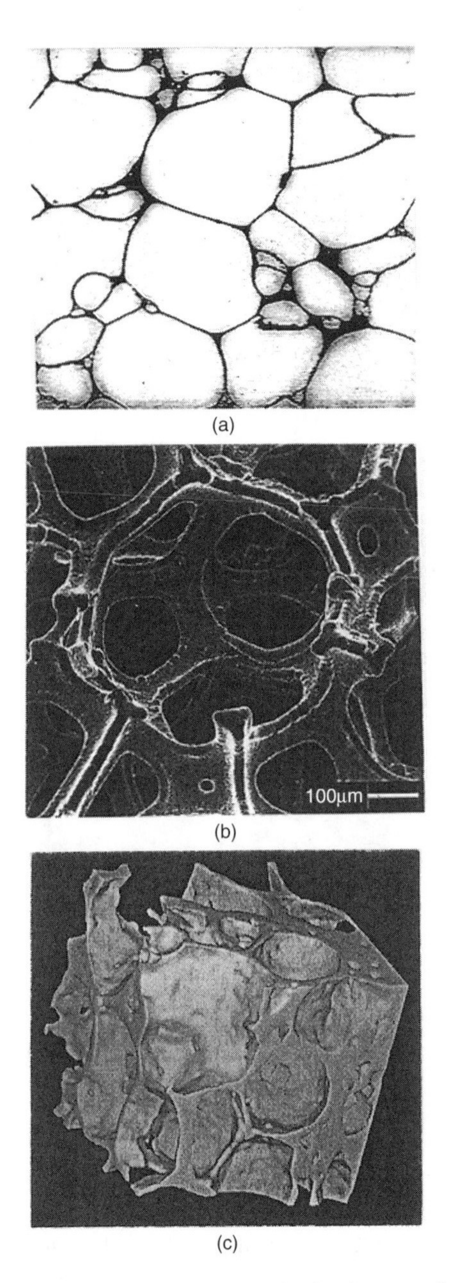

Figure 3.1 *(a) An optical micrograph of a polished section of an Alcan aluminum foam. (b) A SEM micrograph of an INCO nickel foam (Kriszt and Ashby, 1997). (c) An X-ray tomograph of an Alulight foam foam (B Kriszt, private communication, 1999)*

most informative for open-cell foams (Figure 3.1(b)). Closed-cell foams often present a confusing picture from which reliable data for size and shape are not easily extracted. For these, optical microscopy is often better.

X-ray Computed Tomography (CT) gives low magnification images of planes within a foam which can be assembled into a three-dimensional image (Figure 3.1(c)). Medical CT scanners are limited in resolution to about 0.7 mm; industrial CT equipment can achieve 200 μ. The method allows examination of the interior of a closed-cell foam, and is sufficiently rapid that cell distortion can be studied through successive imaging as the sample is deformed.

3.2 Surface preparation and sample size

Metallic foam specimens can be machined using a variety of standard techniques. Cell damage is minimized by cutting with a diamond saw, with an electric discharge machine or by chemical milling. Cutting with a bandsaw gives a more ragged surface, with some damage. The measured values of Young's modulus and compressive strength of a closed-cell aluminum foam cut by diamond-sawing and by electric discharge machining are identical; but the values measured after cutting with a bandsaw are generally slightly lower (Young's modulus was reduced by 15% while compressive strength was reduced by 7%). Thus surface preparation prior to testing or microscopy is important.

The ratio of the specimen size to the cell size can affect the measured mechanical properties of foams (Figure 3.2). In a typical uniaxial compression test, the two ends of the sample are in contact with the loading platens, and the sides of a specimen are free. Cell walls at the sides are obviously less constrained than those in the bulk of the specimen and contribute less to the stiffness and strength. As a result, the measured value of Young's modulus and the compressive strength increases with increasing ratio of specimen size to cell size. As a rule of thumb, boundary effects become negligible if the ratio of the specimen size to the cell size is greater than about 7.

Shear tests on cellular materials are sometimes performed by bonding a long, slender specimen of the test material to two stiff plates and loading the along the diagonal of the specimen (ASTM C-273 – see Figure 3.5, below). Bonding a foam specimen to stiff plates increases the constraint of the cell walls at the boundary, producing a stiffening effect. Experimental measurements on closed-cell aluminum foams, and analysis of geometrically regular, two-dimensional honeycomb-like cellular materials, both indicate that the boundary effects become negligible if the ratio of the specimen size to the cell size is greater than about 3.

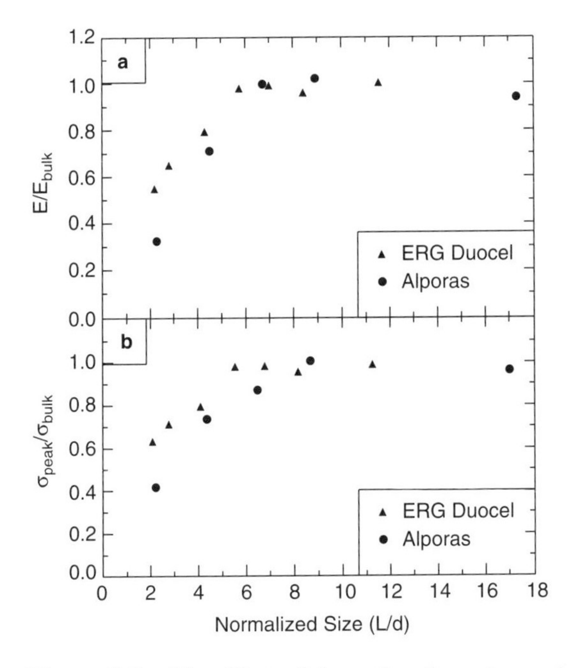

Figure 3.2 *The effect of the ratio of specimen size to cell size on Young's modulus (above) and on compressive plateau stress (below) for two aluminum foams (Andrews et al., 1999b). The modulus and strength become independent of size when the sample dimensions exceed about seven cell diameters*

3.3 Uniaxial compression testing

Uniaxial compressive tests are best performed on prismatic or cylindrical specimens of foam with a height-to-thickness ratio exceeding 1.5. The minimum dimension of the specimen should be at least seven times the cell size to avoid size effects. Displacement can be measured from crosshead displacement, by external LVDTs placed between the loading platens, or by an extensometer mounted directly on the specimen. The last gives the most accurate measurement, since it avoids end effects. In practice, measurements of Young's modulus made with an extensometer are about 5–10% higher than those made using the cross-head displacement.

A typical uniaxial compression stress–strain curve for an aluminum foam is shown in Figure 3.3. The slope of the initial loading portion of the curve is lower than that of the unloading curve. Surface strain measurements (Section 3.10) indicate that there is localized plasticity in the specimen at stresses well below the compressive strength of the foam, reducing the slope of the loading curve. As a result, measurements of Young's modulus should be

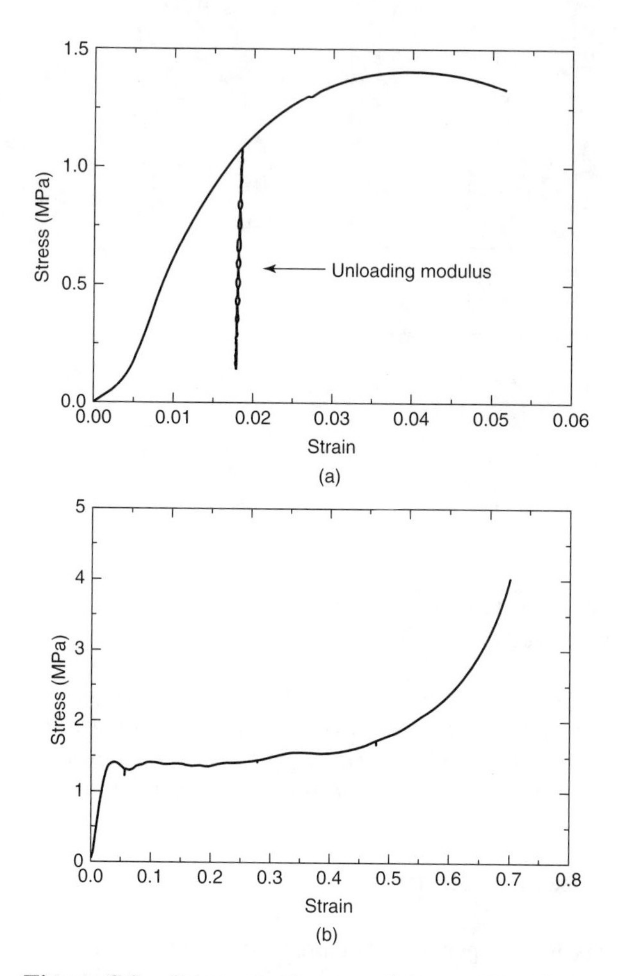

Figure 3.3 *Stress–strain curve from a uniaxial compression test on a cubic specimen of a closed-cell aluminum foam (8% dense Alporas): (a) to 5% strain, (b) to 70% strain (from Andrews et al., 1999a)*

made from the slope of the unloading curve, as shown in Figure 3.3, unloading from about 75% of the compressive strength. The compressive strength of the foam is taken to be the initial peak stress if there is one; otherwise, it is taken to be the stress at the intersection of two slopes: that for the initial loading and that for the stress plateau. Greasing the faces of the specimen in contact with the loading platens reduces frictional effects and can give an apparent compressive strength that is up to 25% higher than that of a dry specimen.

Variations in the microstructure and cell wall properties of some present-day foams gives rise to variability in the measured mechanical properties. The standard deviation in the Young's modulus of aluminum foams is typically between 5% and 30% of the mean while that in the compressive strength is typically between 5% and 15%. Data for the compressive strength of metallic foams are presented in Chapter 4.

3.4 Uniaxial tension testing

Uniaxial tension tests can be performed on either waisted cylinder or dogbone specimens. The specimens should be machined to the shape specified in ASTM E8-96a, to avoid failure of the specimen in the neck region or at the grips. The minimum dimension of the specimen (the diameter of the cylinder or the thickness of the dogbone) should be at least seven times the cell size to avoid specimen/cell size effects. Gripping is achieved by using conventional grips with sandpaper to increase friction, or, better, by adhesive bonding.

Displacement is best measured using an extensometer attached to the waisted region of the specimen. A typical tensile stress–strain curve for an aluminum foam is shown in Figure 3.4. Young's modulus is measured from the unloading portion of the stress–strain curve, as in uniaxial compression testing. The tensile strength is taken as the maximum stress. Tensile failure strains are low for aluminum foams (in the range of 0.2–2%). The standard deviation in the tensile strengths of aluminum foams, like that of the compressive strength, is between 5% and 15% of the mean. Typical data for the tensile strength of metallic foams are given in Chapter 4.

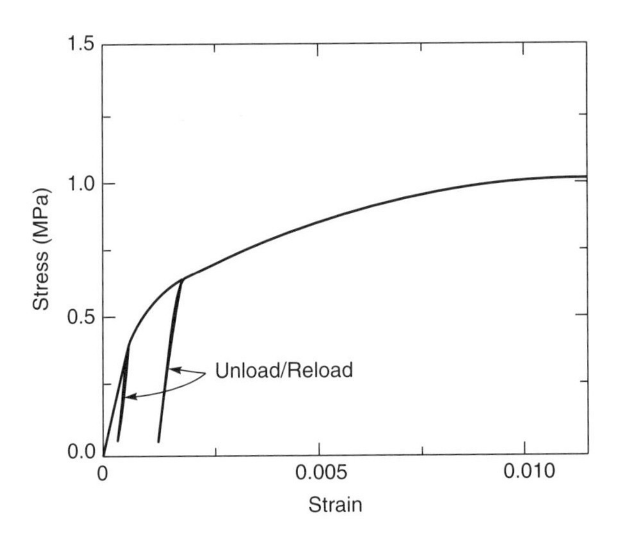

Figure 3.4 *Stress–strain curve from a uniaxial tension test on a dogbone specimen of a closed-cell aluminum foam (8% dense Alporas) (from Andrews et al., 1999a)*

3.5 Shear testing

The shear modulus of metallic foams is most easily measured by torsion tests on waisted cylindrical specimens. The specimens should be machined to the shape of ASTM E8-96a, to avoid failure of the specimen in the neck region or at the grips. The minimum dimension of the specimen (the diameter of the cylinder) should be at least seven times the cell size to avoid specimen/cell size effects. Torque is measured from the load cell. Displacement is measured using two wires, separated by some gauge length. The wires are attached to the specimen at one end, drawn over a pulley and attached to a linear voltage displacement transducer (LVDT) at the other end. The motion of the LVDTs can be converted into the angle of twist of the specimen over the gauge length, allowing the shear modulus to be calculated. The shear modulus is again measured from the unloading portion of the stress–strain curve, as in uniaxial compression testing. The shear strength is taken as the maximum stress. The standard deviation in the shear strengths of aluminum foams are similar those for the compressive and tensile strengths.

A alternative test for measurement of shear strength is ASTM C-273. A long thin specimen is bonded to two stiff plates and the specimen is loaded in tension along the diagonal using commercially available loading fixtures (Figure 3.5(a)). If the specimen is long relative to its thickness (ASTM C-273 specifies $L/t > 12$) then the specimen is loaded in almost pure shear. Metallic

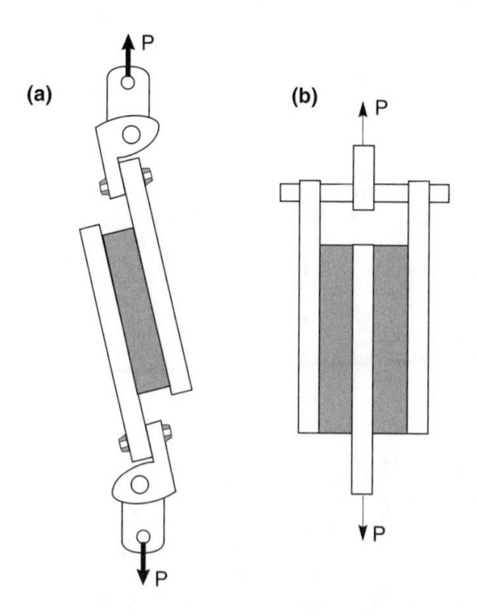

Figure 3.5 *Measurement of shear strength of a foam (a) by the ASTM C-273 test method, and (b) by the double-lap shear test*

foam specimens can be bonded to the plates using a structural adhesive (e.g. FM300 Cytec, Havre de Grace, MD). The load is measured using the load cell while displacement is measured from LVDTs attached to the plates.

The double-lap configureation, shown in Figure 3.5(b), produces a more uniform stress state in the specimen and is preferred for measurement of shear strength, but it is difficult to design plates that are sufficiently stiff to measure the shear modulus reliably. Data for the shear modulus and strength of metallic foams are given in Chapter 4.

3.6 Multi-axial testing of metal foams

A brief description of an established test procedure used to measure the multi-axial properties of metal foams is given below. Details are given in Deshpande and Fleck (2000) and Gioux *et al.* (2000).

Apparatus

A high-pressure triaxial system is used to measure the axisymmetric compressive stress–strain curves and to probe the yield surface. It consists of a pressure cell and a piston rod for the application of axial force, pressurized with hydraulic fluid. A pressure p gives compressive axial and radial stresses of magnitude p. Additional axial load is applied by the piston rod, driven by a screw-driven test frame, such that the total axial stress is $p + \sigma$. The axial load is measured using a load cell internal to the triaxial cell, and the axial displacement is measured with a LVDT on the test machine cross-head and recorded using a computerized data logger. The cylindrical test samples must be large enough to ensure that the specimens have at least seven cells in each direction. The specimens are wrapped in aluminum shim (25 μm thick), encased in a rubber membrane and then sealed using a wedge arrangement as shown in Figure 3.6. This elaborate arrangement is required in order to achieve satisfactory sealing at pressures in excess of 5 MPa.

With this arrangement, the mean stress σ_m and the von Mises effective stress σ_e follow as

$$\sigma_m = -\left(p + \frac{\sigma}{3}\right) \tag{3.1}$$

and

$$\sigma_e = |\sigma| \tag{3.2}$$

respectively. Note that the magnitude of the radial Cauchy stress on the specimen equals the fluid pressure p while the contribution σ to the axial Cauchy stress is evaluated from the applied axial force and the current cross-sectional area of the specimen.

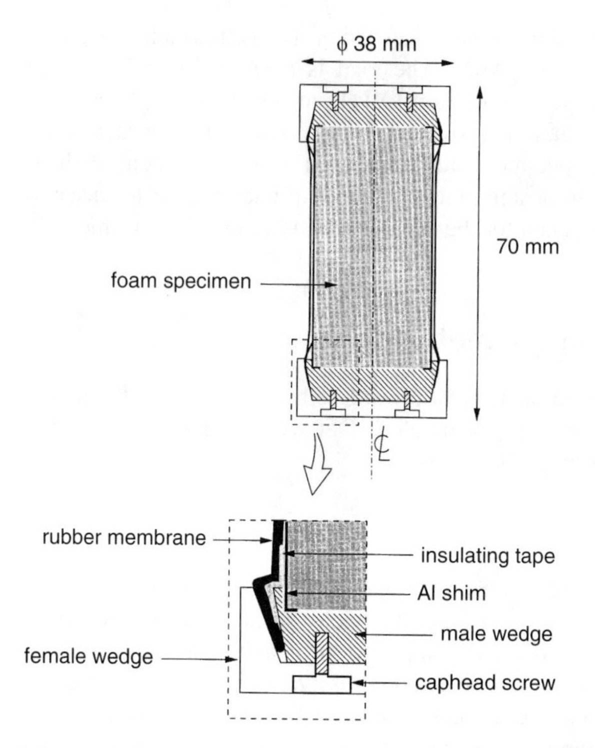

Figure 3.6 *Specimen assembly for multiaxial testing*

The stress–strain curves

Three types of stress versus strain curves are measured as follows:

- Uniaxial compression tests are performed using a standard screw-driven test machine. The load is measured by the load cell of the test machine and the machine platen displacement is used to define the axial strain in the specimen. The loading platens are lubricated with PTFE spray to reduce friction. In order to determine the plastic Poisson's ratio, an essential measurement in establishing the constitutive law for the foam (Chapter 7), the specimens are deformed in increments of approximately 5% axial plastic strain and the diameter is measured at three points along the length of the specimen using a micrometer. The plastic Poisson's ratio is defined as the negative ratio of the transverse to the axial logarithmic strain increment.
- Hydrostatic compression tests are performed increasing the pressure in increments of 0.1 MPa and recording the corresponding volumetric strain, deduced from the axial displacement. The volumetric strain is assumed to be three times the axial strain. *A posteriori* checks of specimen deformation must be performed to confirm that the foams deform in an isotropic manner.

- Proportional axisymmetric stress paths are explored in the following way. The direction of stressing is defined by the relation $\sigma_m = -\eta\sigma_e$, with the parameter taking values over the range $\eta = \frac{1}{3}$ (for uniaxial compression) to $\eta = \infty$ (for hydrostatic compression). In a typical proportional loading experiment, the hydrostatic pressure and the axial load are increased in small increments keeping η constant. The axial displacement are measured at each load increment and are used to define the axial strain.

Yield surface measurements

The initial yield surface for the foam is determined by probing each specimen through the stress path sketched in Figure 3.7. First, the specimen is pressurized until the offset axial plastic strain is 0.3%. This pressure is taken as the yield strength under hydrostatic loading. The pressure is then decreased slightly and an axial displacement is applied until the offset axial strain has incremented by 0.3%. The axial load is then removed and the pressure is decreased further, and the procedure is repeated. This probing procedure is continued until the pressure p is reduced to zero; in this limit the stress state consists of uniaxial compressive axial stress. The locus of yield points, defined at 0.3% offset axial strain, are plotted in mean stress-effective stress space.

In order to measure the evolution of the yield surface under uniaxial loading, the initial yield surface is probed as described above. The specimen is then compressed uniaxially to a desired level of axial strain and the axial load is removed; the yield surface is then re-probed. By repetition of this technique, the evolution of the yield surface under uniaxial loading is measured at a

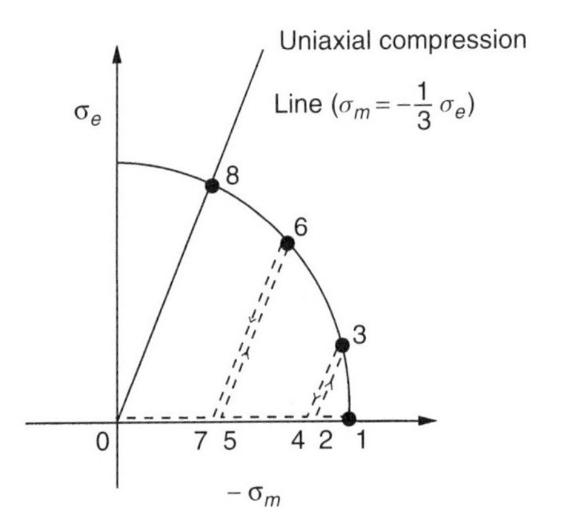

Figure 3.7 *Probing of the yield surface. In the example shown, the specimen is taken through the sequence of loading states 0,1,2,3,4,5,6,7,0,8. The final loading segment 0 → 8 corresponds to uniaxial compression*

number of levels of axial strain from a single specimen. The evolution of the yield surface under hydrostatic loading is measured in a similar manner. Data for the failure of metallic foams under multiaxial loading are described in Chapter 7.

3.7 Fatigue testing

The most useful form of fatigue test is the stress-life S–N test, performed in load control. Care is needed to define the fatigue life of a foam specimen. In tension–tension fatigue and in shear fatigue, there is an incubation period after which the specimen lengthens progressively with increasing fatigue cycles. The failure strain is somewhat less than the monotonic failure strain, and is small; a knife-edge clip gauge is recommended to measure strain in tension or in shear. The fatigue life is defined as the number of cycles up to separation. In compression–compression fatigue there is an incubation period after which the specimen progressively shortens, accumulating large plastic strains of the order of the monotonic lock-up strain. Axial strain is adequately measured by using the cross-head displacement of the test frame. The fatigue life is defined as the number of cycles up to the onset of progressive shortening. As noted in Section 3.2, it is important to perform fatigue tests on specimens of adequate size. As a rule of thumb, the gauge section of the specimen should measure at least seven cell dimensions in each direction, and preferably more.

Compression–compression fatigue is best explored by loading cuboid specimens between flat, lubricated platens. It is important to machine the top and bottom faces of the specimens flat and parallel (for example, by spark erosion) to prevent failure adjacent to the platens. Progressive axial shortening commences at a strain level about equal to the monotonic yield strain for the foam (e.g. 2% for Alporas of relative density 10%). The incubation period for the commencement of shortening defines the fatigue life N_f. The progressive shortening may be uniform throughout the foam or it may be associated with the sequential collapse of rows of cells.

Tension–tension fatigue requires special care in gripping. It is recommended that tests be performed on a dogbone geometry, with cross-sectional area of the waisted portion of about one half that of the gripped ends, to ensure failure remote from the grips. Slipping is prevented by using serrated grips or adhesives.

Shear fatigue utilizes the loading geometries shown in Figures 3.5 and 3.8:

1. The ASTM C273 lap-shear test applied to fatigue (Figure 3.5(a))
2. The double-lap shear test (Figure 3.5(b))
3. The sandwich panel test in the core-shear deformation regime (Figure 3.8)

Initial evidence suggests that the measured S–N curve is insensitive to the particular type of shear test. The ASTM lap-shear test involves large specimens

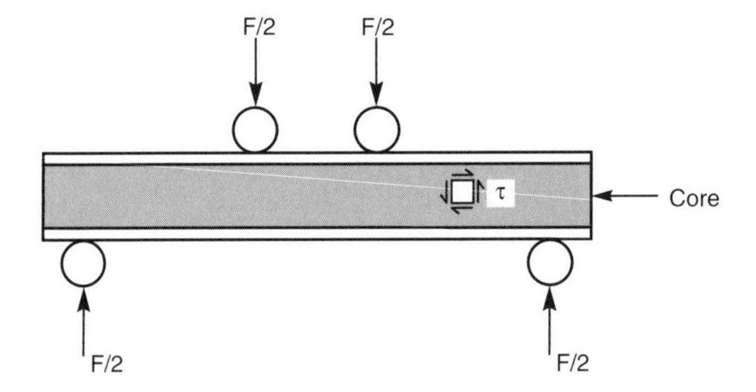

Figure 3.8 *The sandwich-beam test, configured so that the core is loaded predominantly in shear*

which may be difficult to obtain in practice. The sandwich panel test has the virtue that it is closely related to the practical application of foams as the core of a sandwich panel.

3.8 Creep testing

Creep tests are performed using a standard testing frame which applies a dead load via a lever arm to a specimen within a furnace. The temperature of the furnace should be controled to within 1°C. Displacement is measured using an LVDT attached to the creep frame in such a way that although it measures the relative displacement of the specimen ends, it remains outside the furnace. Compression tests are performed by loading the specimen between two alumina platens. Tension tests are performed by bonding the specimens to stainless steel grip pieces with an aluminum oxide-based cement (e.g. Sauereisen, Pittsburg, PA). The steel pieces have holes drilled through them and the cement is forced into the holes for improved anchorage.

Data for creep of aluminum foams can be found in Chapter 9.

3.9 Indentation and hardness testing

Reproducible hardness data require that the indenter (a sphere or a flat-ended cylinder) have a diameter, D, that is large compared with the cell size, $d(D/d > 7)$. Edge effects are avoided if the foam plate is at least two indenter diameters in thickness and if the indentations are at least one indenter diameter away from the edges of the plate. Because they are compressible, the indentation strength of a foam is only slightly larger than its uniaxial compressive strength. By contrast, a fully dense solid, in which volume is conserved during plastic deformation, has a hardness that is about three times its yield strength for shallow indentation and about five times for deep indentation. The indentation strength of foams is slightly larger than the uniaxial compressive strength because of the work done in tearing cell walls around the perimeter of the indenter and because of friction, but these contributions diminish as the indent diameter increases.

3.10 Surface strain mapping

The strain field on the surface of a metallic foam resulting from thermomechanical loading can be measured using a technique known as *surface strain mapping*. The surfaces of cellular metals are irregular, with the cell membranes appearing as peaks and troughs, allowing *in-situ* optical imaging to be used to provide a map of surface deformation. Commercial surface displacement analysis equipment and software (SDA) are available from Instron (1997). The SDA software performs an image correlation analysis by comparing pairs of digital images captured during the loading history. The images are divided into sub-images, which provide an array of analysis sites across the surface. Displacement vectors from these sites are found by using 2D-Fast Fourier Transform (FFT) comparisons of consecutive pairs of sub-images.

The method requires surface imaging, for which a commercial video camera with a CCD array of 640×480 or 1024×1528 pixels is adequate, preferably with a wide-aperture lens (F/1.4) and fiber-optic light source. Since cellular metals exhibit non-uniform, heterogeneous deformation, the field of view should be optimized such that each unit cell can be mapped to approximately 50 pixels in each direction. The analysis can be carried out by applying FFTs to a 32-pixel square array of sub-images, centered at nodal points eight pixels apart, such that the deformation of each unit cell is represented by at least four nodal points in each direction.

The method relies on the recognition of surface pattern. The foam surface can be imaged directly, relying on the irregular pattern of surface cell-edges for matching between consecutive frames. Alternatively, a pre-stretched latex film sprayed with black and white emulsion to give a random pattern can be bonded to the surface. During loading, the film follows the cell shape changes without delamination. While the latex film method is more accurate, direct imaging of the surface provides essentially the same continuum deformation field, and is preferred because of its simplicity.

Deformation histories for the Alporas material are visualized as false color plots of components of strain in the plane of the surface (Figure 3.9). Maps of the incremental distortion at loadings between the start of the non-linear response and the onset of the plateau reveal that localized deformation bands initiate at the onset of non-linearity having width about one cell diameter. Within each band, there are cell-sized regions that exhibit strain levels about

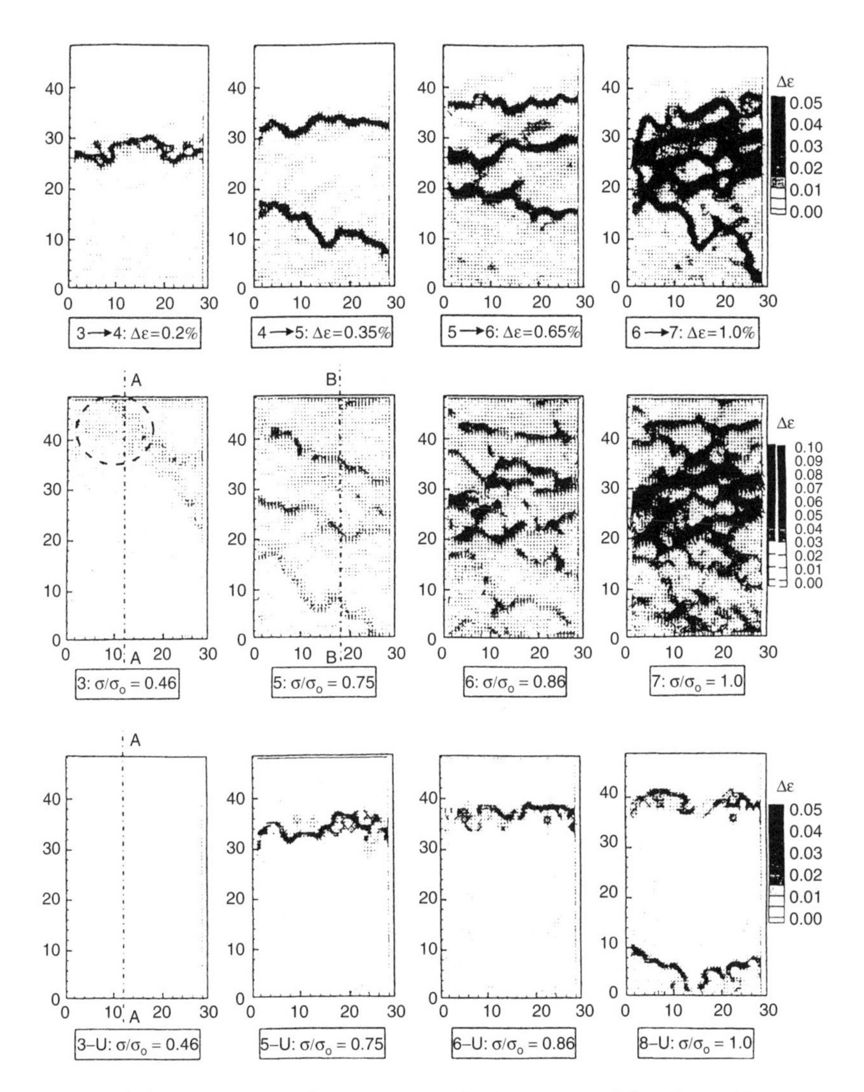

Figure 3.9 *Distortional strain maps for incremental loading : Top row: incremental distortion at various load levels. Middle row: maps of accumulated distortion at various load levels along the deformation history. Bottom row: incremental distortion at various unloading levels*

an order of magnitude larger than the applied strain. Outside the bands, the average strains are small and within the elastic range. The principal strains reveal that the flow vectors are primarily in the loading direction, normal to the band plane, indicative of a crushing mode of deformation. The cumulative distortions exhibit similar effects over the same strain range.

As an example or the information contained in surface strain maps, consider the following features of Figure 3.9:

1. Strain is non-uniform, as seen in the top set of images. Bands form at the onset of non-linearity (site A) and then become essentially inactive. Upon further straining new bands develop. Some originate at previously formed bands, while others appear in spatially disconnected regions of the gage area.
2. Deformation starts at stress levels far below general yield. Plasticity is evident in the second set of images at stresses as low as 0.45 of the plateau stress.
3. Some of the strain is reversible. The bands in the bottom sequence of images show reverse straining as the sample is unloaded.

3.11 Literature on testing of metal foams

Structural characterization

Bart-Smith, H., Bastawros, A.-F., Mumm, D.R., Evans, A.G., Sypeck, D. and Wadley, H.G. (1998) Compressive deformation and yielding mechanisms in cellular Al alloys determined Using X-ray tomography and surface strain mapping. *Acta Mater.* **46**: 3583–3592.

Kottar, A., Kriszt, B. and Degisher, H.P. (1999) Private communication.

Mechanical testing

American Society for Testing and Materials Specification C-273-61, Shear test in flatwise plane of flat sandwich constructions or sandwich cores. American Society for Testing and Materials, Philadelphia, PA.

American Society for Testing and Materials Specification E8-96a, Standard test methods for tension testing of metallic materials. American Society for Testing and Materials, Philadelphia, PA.

Andrews, E.W., Sanders, W. and Gibson, L.J. (1999a) Compressive and tensile behavior of aluminum foams. *Materials Science and Engineering* **A270**, 113–124.

Andrews, E.W., Gioux, G., Onck, P. and Gibson, L.J. (1999b) Size effects in ductile cellular solids, submitted to *Int. J. Mech. Sci.*

Strain mapping

Bastawros, A.-F. and McManuis, R. (1998) Case study: Use of digital image analysis software to measure non-uniform deformation in cellular aluminum alloys. *Exp. Tech.* **22**, 35–37.

Brigham, E.O. (1974) *The Fast Fourier Transform*, Prentice-Hall, Englewood Cliffs, NJ.

Chen, D.J., Chiang, F.P., Tan, Y.S. and Don, H.S. (1993) Digital speckle-displacement measurement using a complex spectrum method. *Applied Optics*. **32**, 1839–1849.

Instron, (1997) *Surface Displacement Analysis User Manual.*

Multiaxial testing of foams

Deshpande, V.S. and Fleck, N.A. (2000) Isotropic constitutive models for metallic foams. To appear in *J. Mech. Phys. Solids*.

Gioux, G., McCormick, T. and Gibson, L.J. (2000) Failure of aluminum foams under multi-axial loads. *Int. J. Mech. Sci.* **42**, 1097–1117.

Hardness testing of foams

Shaw, M.C. and Sata, T. (1966) *Int. J. Mech. Sci.* **8**, 469.
Wilsea, M., Johnson, K.L. and Ashby, M.F. (1975) *Int. J. Mech. Sci.* **17**, 457.

Chapter 4

Properties of metal foams

The characteristics of a foam are best summarized by describing the material from which it is made, its relative density, ρ/ρ_s (the foam density, ρ, divided by that of the solid material of the cell wall, ρ_s), and stating whether it has open or closed cells. Beyond this, foam properties are influenced by structure, particularly by anisotropy and by defects – by which we mean wiggly, buckled or broken cell walls, and cells of exceptional size or shape.

Metal foams are still inadequately characterized, but the picture is changing rapidly. An overview of the range spanned by their properties is given by Table 4.1 and the property charts of Section 4.3. The primary links between properties, density and structure are captured in scaling relations, listed in Section 4.4. They allow foam properties to be estimated, at an approximate level, when solid properties are known.

The producers of metal foams have aggressive development programs for their materials. The properties described here are those of the currently available generation of foams, and should be regarded as a basis for initial, scoping, calculations and designs. The next generation of foams will certainly be better. Final design calculations must be based on data provided by the supplier.

4.1 Foam structure

Figures 4.1(a)–(c) show the structure of metal foams from three different suppliers: Cymat, Mepura (Alulight) and Shinko (Alporas). The structures are very like those of soap films: polyhedral cells with thin cell faces bordered by thicker cell edges ('Plateau borders'). Some of the features appear to be governed by surface energy, as they are in soap films: the Plateau borders are an example. But others are not: many faces have non-uniform curvature or are corrugated, and have occasional broken walls that still hang in place.

The three figures are ordered such that the relative density increases from the top to the bottom. The Cymat (Al–SiC) foam in Figure 4.4(a) has a relative density $\rho/\rho_s = 0.05$, and an average cell size approaching 5 mm; foams from this source are available in the range $0.02 < \rho/\rho_s < 0.2$. The Alporas (Al–Ca) foam in Figure 4.4(b) has smaller cells and comes in a narrower range of relative density: $0.08 < \rho/\rho_s < 0.2$; that shown here has a value of

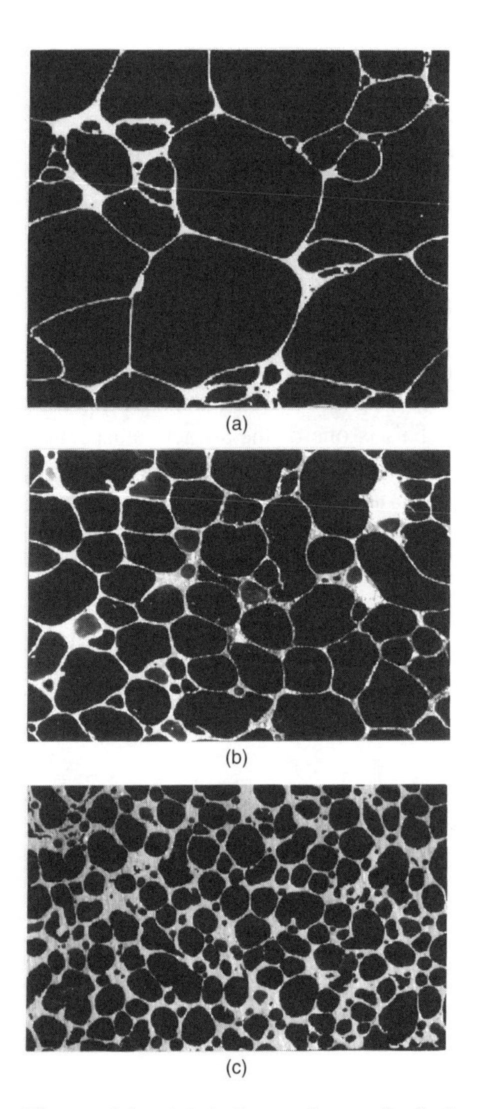

Figure 4.1 *(a) A Cymat foam of relative density $\rho/\rho_s = 0.04$ (density 108 kg/m^3 or 6.7 lb ft^3). (b) An Alporas foam of relative density $\rho/\rho_s = 0.09$ (density 240 kg/m^3 or 15 lb ft^3). (c) An Alulight foam of relative density $\rho/\rho_s = 0.25$ (density 435 kg/m^3 or 270 lb ft^3)*

0.09. The Alulight foam (Al–TiH) in Figure 4.4(c) has a relative density of 0.25, which lies at the upper end of the range in which this material is made ($0.1 < \rho/\rho_s < 0.35$).

The properties of metal foams depend most directly on those of the material from which they are made and on their relative density; but they are influenced by structure too. This influence, imperfectly understood at present, is a topic of intense current study. Better understanding will lead to greater process control and improved properties. But for now we must accept the structures as they are, and explore how best to design with the present generation of foams. With this in mind, we document here the properties of currently available, commercial, metal foams.

4.2 Foam properties: an overview

The ranges of properties offered by currently available metal foams are documented in Table 4.1. Many suppliers offer a variety of densities; the properties, correspondingly, exhibit a wide range. This is one of the attractive aspects of such materials: a desired profile of properties can be had by selecting the appropriate foam material with the appropriate density.

Nomenclature and designation

It is important to distinguish between the properties of the metfoam and those of the solid from which it is made. Throughout this Guide, properties subscripted with s refer to the solid from which the foam is made (e.g. solid density: ρ_s); properties without the subscript s are those of the metfoam (foam density: ρ).

Data sources

Data can be found in the literature cited in Section 1.4, and in recent publications in leading materials journals. The most comprehensive data source is the *CES* (1999) database and associated literature; it draws on manufacturers' data sheets, on published literature, and on data supplied by research groups, worldwide (Evans, 1997/1998/1999; Banhart, 1999; Degisher, 1999; Simancik, 1999). The *CES* database has been used to create the material property charts shown in Section 4.3 and later chapters.

Mechanical properties

Figures 4.2 and 4.3 show a schematic stress–strain curve for compression, together with two sets of real ones. Initial loading apears to be elastic but the initial loading curve is not straight, and its slope is less than the true modulus, because some cells yield at very low loads. The real *modulus* E, is best measured dynamically or by loading the foam into the plastic range, then unloading and determining E from the unloading slope. Young's modulus, E,

Table 4.1

(a) Ranges[a] for mechanical properties of commercial[b] metfoams

Property, (units), symbol	*Cymat*	*Alulight*	*Alporas*	*ERG*	*Inco*
Material	Al–SiC	Al	Al	Al	Ni
Relative density (–), ρ/ρ_s	0.02–0.2	0.1–0.35	0.08–0.1	0.05–0.1	0.03–0.04
Structure (–)	Closed cell	Closed cell	Closed cell	Open cell	Open cell
Density (Mg/m^3), ρ	0.07–0.56	0.3–1.0	0.2–0.25	0.16–0.25	0.26–0.37
Young's modulus (GPa), E	0.02–2.0	1.7–12	0.4–1.0	0.06–0.3	0.4–1.0
Shear modulus (GPa), G	0.001–1.0	0.6–5.2	0.3–0.35	0.02–0.1	0.17–0.37
Bulk modulus (GPa), K	0.02–3.2	1.8–13.0	0.9–1.2	0.06–0.3	0.4–1.0
Flexural modulus (GPa), E_f	0.03–3.3	1.7–12.0	0.9–1.2	0.06–0.3	0.4–1.0
Poisson's ratio (–), ν	0.31–0.34	0.31–0.34	0.31–0.34	0.31–0.34	0.31–0.34
Comp. strength (MPa), σ_c	0.04–7.0	1.9–14.0	1.3–1.7	0.9–3.0	0.6–1.1
Tensile elastic limit (MPa), σ_y	0.04–7.0	2.0–20	1.6–1.8	0.9–2.7	0.6–1.1
Tensile strength (MPa), σ_t	0.05–8.5	2.2–30	1.6–1.9	1.9–3.5	1.0–2.4
MOR (MPa), σ_{MOR}	0.04–7.2	1.9–25	1.8–1.9	0.9–2.9	0.6–1.1
Endurance limit (MPa), σ_e^c	0.02–3.6	0.95–13	0 9–1.0	0.45–1.5	0.3–0.6
Densification strain (–), ε_D	0.6–0.9	0.4–0.8	0.7–0.82	0.8–0.9	0.9–0.94
Tensile ductility (–), ε_f	0.01–0.02	0.002–0.04	0.01–0.06	0.1–0.2	0.03–0.1
Loss coefficient (%), η^c	0.4–1.2	0.3–0.5	0.9–1.0	0.3–0.5	1.0–2.0
Hardness (MPa), H	0.05–10	2.4–35	2.0–2.2	2.0–3.5	0.6–1.0
Fr. tough. (MPa.m$^{1/2}$), K_{IC}^c	0.03–0.5	0.3–1.6	0.1–0.9	0.1–0.28	0.6–1.0

Table 4.1 *(continued)*

(b) Ranges[a] for thermal properties of commercial[b] metfoams

Property (units), symbol	*Cymat*	*Alulight*	*Alporas*	*ERG*	*Inco*
Material	Al–SiC	Al	Al	Al	Ni
Relative density (–)	0.02–0.2	0.1–0.35	0.08–0.1	0.05–0.1	0.03–0.04
Structure	Closed cell	Closed cell	Closed cell	Open cell	Open cell
Melting point (K), T_m	830–910	840–850	910–920	830–920	1700–1720
Max. service temp. (K), T_{max}	500–530	400–430	400–420	380–420	550–650
Min. service temp. (K), T_{min}	1–2	1–2	1–2	1–2	1–2
Specific heat (J/kg.K), C_p	830–870	910–920	830–870	850–950	450–460
Thermal cond. (W/m.K), λ	0.3–10	3.0–35	3.5–4.5	6.0–11	0.2–0.3
Thermal exp. (10^{-6}/K), α	19–21	19–23	21–23	22–24	12–14
Latent heat, melting (kJ/kg), L	355–385	380–390	370–380	380–395	280–310

(c) Ranges[a] for electrical resistivity of commercial[b] metfoams

Property (units), symbol	*Cymat*	*Alulight*	*Alporas*	*ERG*	*Inco*
Material	Al–SiC	Al	Al	Al	Ni
Relative density (–)	0.02–0.2	0.1–0.35	0.08–0.1	0.05–0.1	0.03–0.04
Structure	Closed cell	Closed cell	Closed cell	Open cell	Open cell
Resistivity (10^{-8} ohm.m), R	90–3000	20–200	210–250	180–450	300–500

[a]The data show the range of properties associated with the range of relative density listed in the third row of the table. The lower values of a property are associated with the lower densities and vica versa, except for densification strain, where the reverse is true.
[b]Contact information for suppliers can be found in Chapter 18.
[c]Data for endurance limit, loss coefficient and fracture toughness must, for the present, be regarded as estimates.

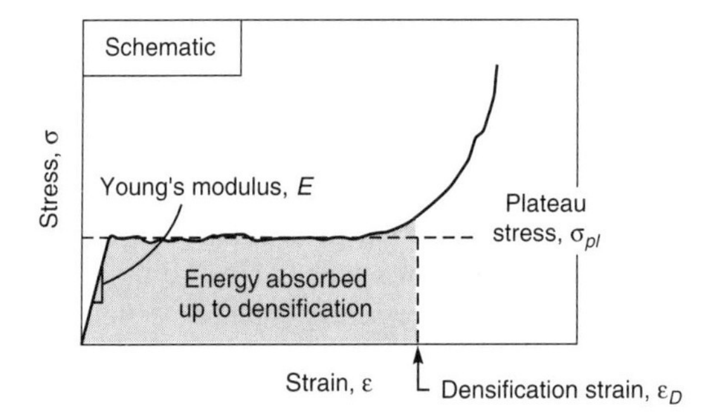

Figure 4.2 *Compression curve for a metal foam – schematic showing properties*

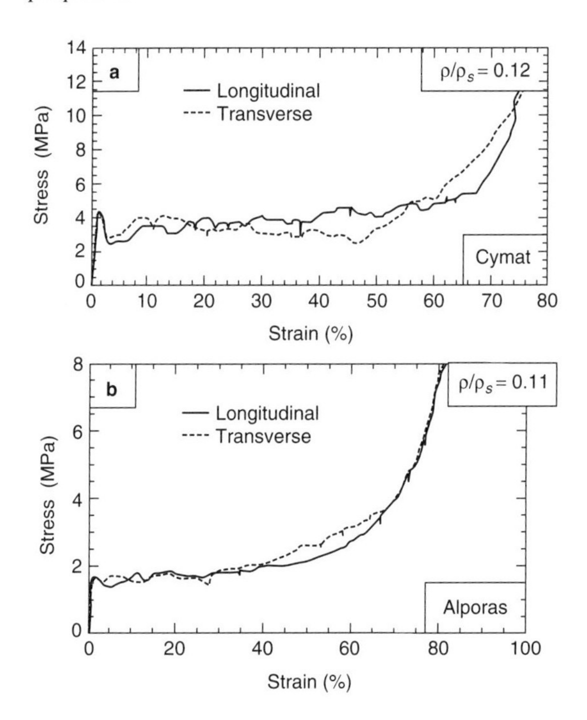

Figure 4.3 *Compression curves for Cymat and Alporas foam*

the shear modulus, G, and Poisson's ratio ν scale with density as:

$$E \approx \alpha_2 E_s \left(\frac{\rho}{\rho_s}\right)^n \quad G \approx \frac{3}{8}\alpha_2 G_s \left(\frac{\rho}{\rho_s}\right)^n \quad \nu \approx 0.3 \tag{4.1}$$

where n has a value between 1.8 and 2.2 and α_2 between 0.1 and 4 – they depend on the structure of the metfoam. As a rule of thumb, $n \approx 2$. For design purposes, it is helpful to know that the tensile modulus E_t of metal foams is not the same as that in compression E_c; the tensile modulus is greater, typically by 10%. Anisotropy of cell shape can lead to significant (30%) differences between moduli in different directions.

Open-cells foams have a long, well-defined *plateau stress*, σ_{pl}, visible on Figures 4.2 and 4.3. Here the cell edges are yielding in bending. Closed-cell foams show somewhat more complicated behavior which can cause the stress to rise with increasing strain because the cell faces carry membrane (tensile) stresses. The plateau continues up to the *densification strain*, ε_D, beyond which the structure compacts and the stress rises steeply. The plateau stress, σ_{pl}, and the densification strain, ε_D, scale with density as:

$$\sigma_{pl} \approx (0.25 \text{ to } 0.35)\sigma_{y,s}\left(\frac{\rho}{\rho_s}\right)^m \qquad \varepsilon_D \approx \left(1 - \alpha_1\frac{\rho}{\rho_s}\right) \tag{4.2}$$

For currently available metfoams m lies between 1.5 and 2.0 and α_1 between 1.4 and 2. As a rule of thumb, $m \approx 1.6$ and $\alpha_1 \approx 1.5$. These properties are important in *energy-absorbing* applications, to which metal foams lend themselves well (see Chapter 11).

The tensile stress–strain behavior of metal foams differs from that in compression. Figure 4.4 shows examples. The slope of the stress–strain curve before general yield is less than E, implying considerable micro-plasticity even at very small strains. Beyond yield (yield strength: σ_y) metal foams harden up to the ultimate tensile strength σ_{ts} beyond which they fail at a tensile ductility ε_t.

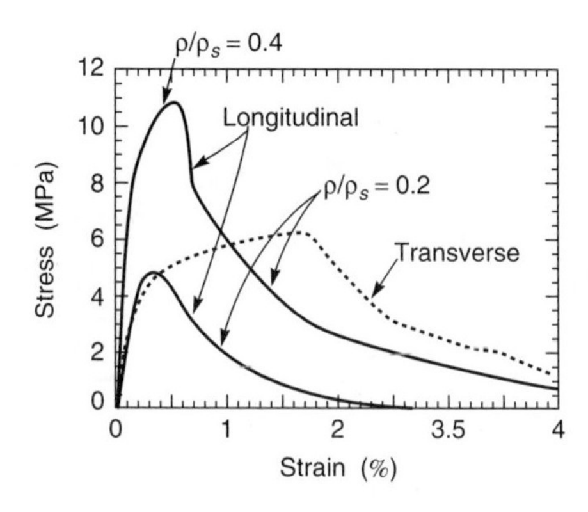

Figure 4.4 *Tensile stress–strain curves for Alulight foams*

The damping capacity of a metal foam is typically five to ten times greater than that of the metal from which it is made. This increase may be useful, although the loss factor is still much less than that associated with polymer foams. Metal foams have some capacity as acoustic absorbers (Chapter 12), although polymer foams and glass wool are generally better.

As in other materials, cyclic loading causes fatigue damage in metal foams. High-cycle fatigue tests allow a fatigue limit $\Delta\sigma_e$ to be measured ($\Delta\sigma_e$ is the cyclic stress range at which the material will just survive 10^7 cycles) Typical data for compression-compression fatigue with an R-value of about 0.1 are shown in Figure 4.5. A detailed description of fatigue behavior of metal foams is given in Chapter 8.

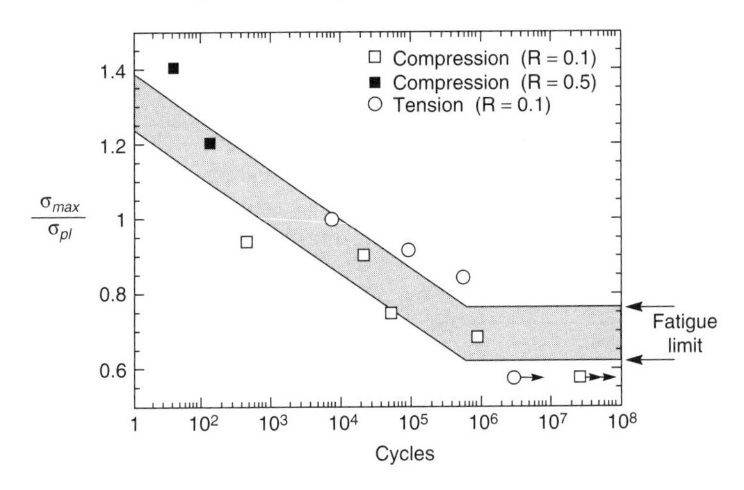

Figure 4.5 *Fatigue data for Alporas foams. Chapter 8 gives details*

The toughness of metal foams can be measured by standard techniques. As a rule of thumb, the initiation toughness J_{IC} scales with density as:

$$J_{IC} \approx \beta\sigma_{y,s} \cdot \ell \left(\frac{\rho}{\rho_s}\right)^p \tag{4.3}$$

where ℓ is the cell size with $p = 1.3$ to 1.5 and $\beta = 0.1$ to 0.4.

The creep of metal foams has not yet been extensively studied. The theory and limited experimental data are reviewed in Chapter 9.

Thermal properties

The melting point, specific heat and expansion coefficient of metal foams are the same as those of the metal from which they are made. The thermal conductivity λ scales with density approximately as:

$$\lambda \approx \lambda_s \left(\frac{\rho}{\rho_s}\right)^q \tag{4.4}$$

with $q = 1.65$ to 1.8.

Electrical properties

The only electrical property of interest is the resistivity, R. This scales with relative density approximately as

$$R \approx R_s \left(\frac{\rho}{\rho_s}\right)^{-r} \tag{4.5}$$

with $r = 1.6 - 1.85$.

4.3 Foam property charts

Figures 4.6–4.11 are examples of material property charts. They give an overview of the properties of metal foams, allow scaling relations to be deduced and enable selection through the use of material indices (Chapter 5). All the charts in this Guide were constructed using the *CES* (1999) software and database.

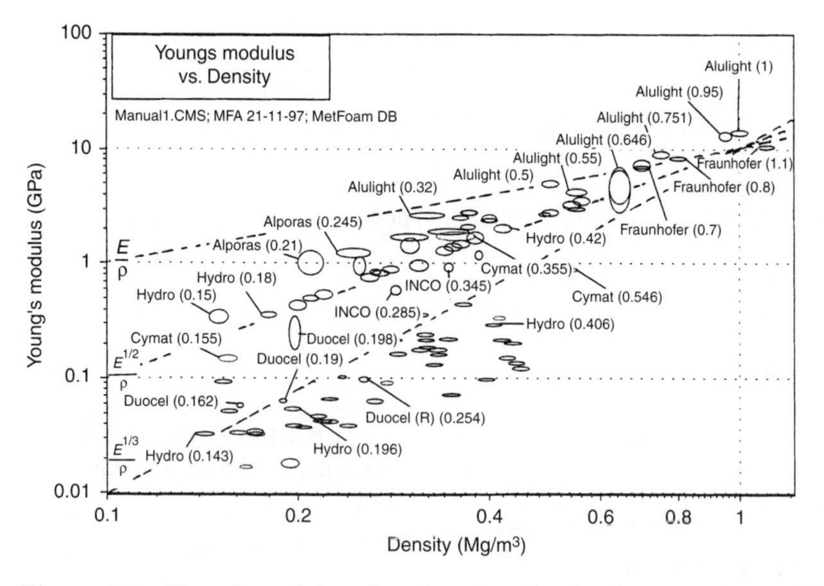

Figure 4.6 *Young's modulus plotted against density for currently available metal foams. Output from CES3.1 with the MetFoam '97 database*

Stiffness and density

Figure 4.6 shows Young's modulus, E, plotted against density ρ for available metal foams. For clarity, only some of the data have been identified. The numbers in parentheses are the foam density in Mg/m^3. The broken lines show the indices E/ρ, $E^{1/2}/\rho$ and $E^{1/3}/\rho$. Their significance is discussed in Chapter 5. Metal foams have attractively high values of the last of these indices, suggesting their use as light, stiff panels, and as a way of increasing natural vibration frequencies.

Strength and density

Figure 4.7 shows compressive strength, σ_c, plotted against density, ρ^* for currently available metal foams. For clarity, only some of the data have been identified. The numbers in parentheses are the foam density in Mg/m^3. The broken lines show the indices σ_c/ρ, $\sigma_c^{2/3}/\rho$ and $\sigma_c^{1/2}/\rho$. Their significance is discussed in Chapter 5. Metal foams have attractively high values of the last of these indices, suggesting their use as light, strong panels.

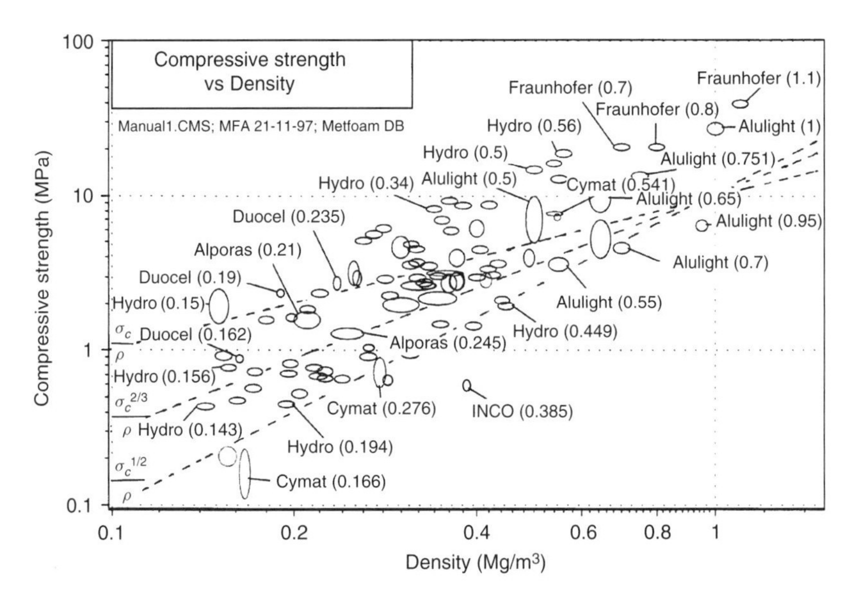

Figure 4.7 *The compressive strength plotted against density for currently available metal foams. Output from CES3.1 with the MetFoam '97 database*

Specific stiffness and strength

Stiffness and strength at low weight are sought in many applications. Caution must be exercised here. If axial stiffness and strength are what is wanted, the proper measure of the first is E/ρ and of the second is σ_c/ρ. But if bending stiffness and strength are sought then $E^{1/2}/\rho$ and $\sigma_c^{2/3}/\rho$ (beams) or $E^{1/3}/\rho$ and $\sigma_c^{1/2}/\rho$ (panels) are the proper measures.

Figure 4.8 shows the first of these combinations. For reference, the value of E/ρ for structural steel, in the units shown here, is 25 GPa/(Mg/m^3), and that for σ_c/ρ is 24 MPa/(Mg/m^3). The values for the 1000 series aluminum alloys are almost the same. Metal foams are have lower values of these two properties than do steel and aluminum. Figure 4.9 shows $E^{1/2}/\rho$ plotted against $\sigma_c^{2/3}/\rho$. Values for steel are 1.8 and 4.3; for aluminum, 3.1 and 6.2, all in the units shown on the figure. Metal foams can surpass conventional materials here. Figure 4.10 shows $E^{1/3}/\rho$ plotted against $\sigma_c^{1/2}/\rho$. Values for steel are 0.7 and 1.8; for aluminum, 1.5 and 3.7, all in the units shown on the figure. Metal foams easily surpass conventional materials in these properties.

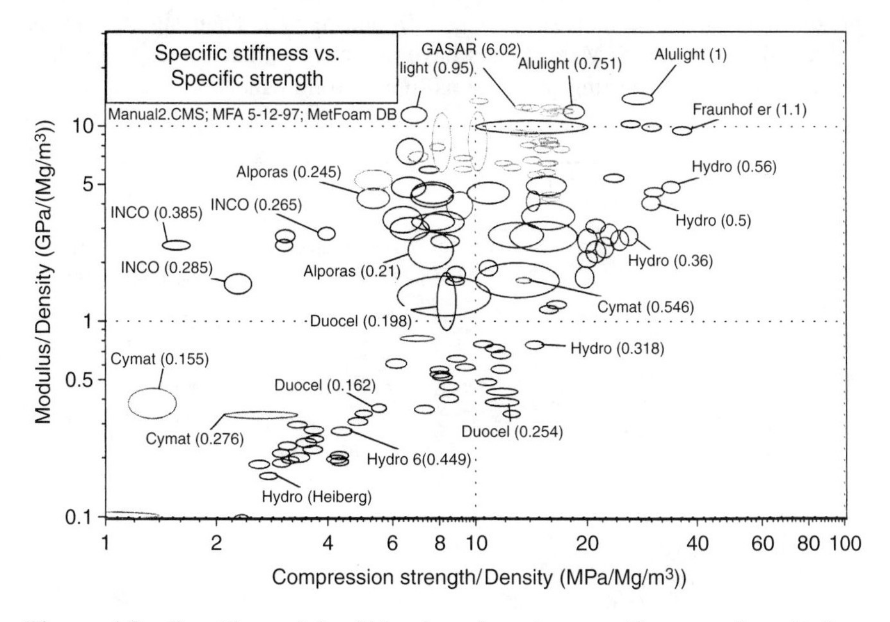

Figure 4.8 *Specific modulus E/ρ plotted against specific strength σ_c/ρ for currently available metal foams*

Thermal properties

Figure 4.11 captures a great deal of information about thermal properties. As before, only some of the data have been identified for clarity. The numbers in parentheses are the foam density in Mg/m^3. The figure shows the thermal conductivity, λ, plotted against the specific heat per unit volume, $C_p\rho$. To this can be added contours of thermal diffusivity

$$a = \frac{\lambda}{C_p \rho}$$

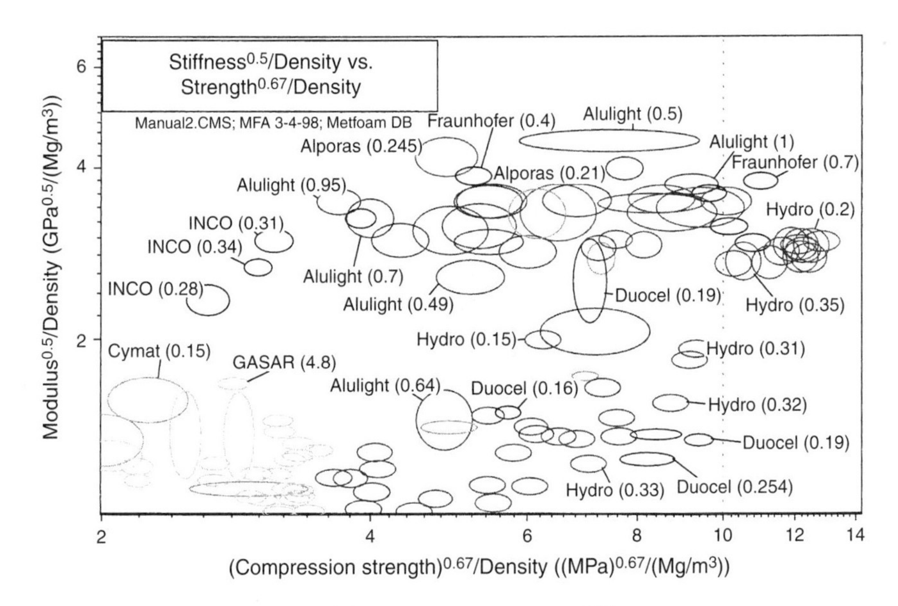

Figure 4.9 *The quantity $E^{1/2}/\rho$ plotted against $\sigma_c^{2/3}/\rho$ for currently available metal foams*

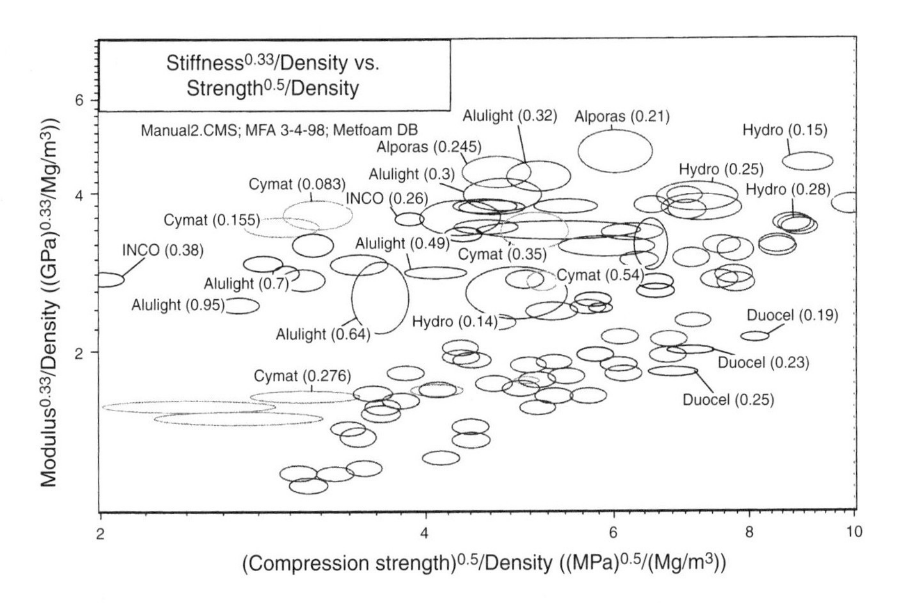

Figure 4.10 *The quantity $E^{1/3}/\rho$ plotted against $\sigma_c^{1/2}/\rho$ for currently available metal foams*

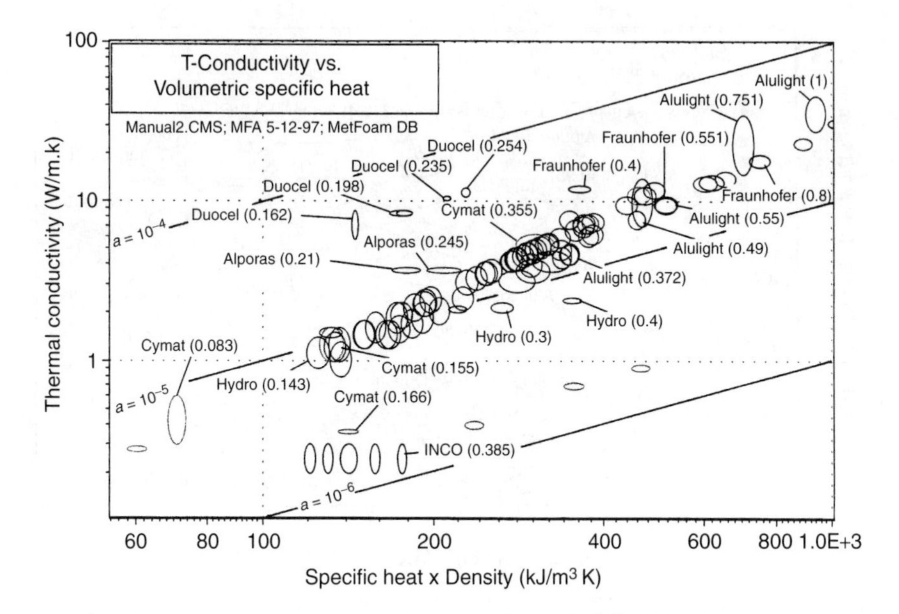

Figure 4.11 *Thermal conductivity, λ, plotted against volumetric specific heat $C_p\rho$ for currently available metal foams. Contours show the thermal diffusivity $a = \lambda/C_p\rho$ in units of m^2/s*

They are shown as full lines on the chart. Thermal conductivity is the property which determines steady-state heat response; thermal diffusivity determines transient response. Foams are remarkable for having low values of thermal conductivity. Note that the open-cell aluminum foams have relatively high thermal diffusivities; the nickel foams have low ones. All closed-cell aluminum foams have almost the same value of *a*.

4.4 Scaling relations

Tables 4.2(a)–(c) give scaling relations for foam properties. They derive partly from modeling (for which see Gibson and Ashby, 1997) most of it extensively tested on polymeric foams, and partly from empirical fits to experimental data of the type shown in the previous section.

The properties are defined in the first column. In the rest of the table a symbol with a subscripted *s* means 'property of the solid metal of which the foam is made'; a symbol with a superscript '*' is a property of the foam. The relations take the form

$$\frac{P^*}{P_s} = \alpha \left\{ \frac{\rho^*}{\rho_s} \right\}^n$$

where P is a property, α a constant and n a fixed exponent.

The scaling relations are particularly useful in the early stages of design when approximate analysis of components and structures is needed to decide

Table 4.2

(a) Scaling laws for the mechanical properties of foams

Mechanical properties	*Open-cell foam*	*Closed-cell foam*
Young's modulus (GPa), E	$E = (0.1-4)E_s \left(\frac{\rho}{\rho_s}\right)^2$	$E = (0.1-1.0)E_s \times \left[0.5\left(\frac{\rho}{\rho_s}\right)^2 + 0.3\left(\frac{\rho}{\rho_s}\right)\right]$
Shear modulus (GPa), G	$G \approx \frac{3}{8}E$	$G \approx \frac{3}{8}E$
Bulk modulus (GPa), K	$K \approx 1.1E$	$K \approx 1.1E$
Flexural modulus (GPa), E_f	$E_f \approx E$	$E_f \approx E$
Poisson's ratio ν	0.32–0.34	0.32–0.34
Compressive strength (MPa), σ_c	$\sigma_c = (0.1-1.0)\sigma_{c,s} \left(\frac{\rho}{\rho_s}\right)^{3/2}$	$\sigma_c = (0.1-1.0)\sigma_{c,s} \times \left[0.5\left(\frac{\rho}{\rho_s}\right)^{2/3} + 0.3\left(\frac{\rho}{\rho_s}\right)\right]$
Tensile strength (MPa), σ_t	$\sigma_t \approx (1.1-1.4)\sigma_c$	$\sigma_t \approx (1.1-1.4)\sigma_c$
Endurance limit (MPa), σ_e	$\sigma_e \approx (0.5-0.75)\sigma_c$	$\sigma_e \approx (0.5-0.75)\sigma_c$
Densification strain, ε_D	$\varepsilon_D = (0.9-1.0) \times \left(1-1.4\frac{\rho}{\rho_s} + 0.4\left(\frac{\rho}{\rho_s}\right)^3\right)$	$\varepsilon_D = (0.9-1.0) \times \left(1-1.4\frac{\rho}{\rho_s} + 0.4\left(\frac{\rho}{\rho_s}\right)^3\right)$
Loss coefficient, η	$\eta \approx (0.95-1.05) \times \frac{\eta_s}{(\rho/\rho_s)}$	$\eta \approx (0.95-1.05) \times \frac{\eta_s}{(\rho/\rho_s)}$
Hardness (MPa), H	$H = \sigma_c\left(1 + 2\frac{\rho}{\rho_s}\right)$	$H = \sigma_c\left(1 + 2\frac{\rho}{\rho_s}\right)$
Initiation toughness. (J/m^2) J_{IC}	$J_{IC}^* \approx \beta\sigma_{y,s}\ell\left(\frac{\rho^*}{\rho_s}\right)^p$	$J_{IC}^* \approx \beta\sigma_{y,s}\ell\left(\frac{\rho^*}{\rho_s}\right)^p$

Table 4.2 *(continued)*

(b) Scaling laws for thermal properties

Thermal properties	*Open-cell foam*	*Closed-cell foam*
Melting point (K), T_m	As solid	As solid
Max. service temp. (K), T_{max}	As solid	As solid
Min. service temp. (K), T_{min}	As solid	As solid
Specific heat (J/kg.K), C_p	As solid	As solid
Thermal cond. (W/m.K), λ	$\left(\frac{\rho}{\rho_s}\right)^{1.8} < \frac{\lambda}{\lambda_s} < \left(\frac{\rho}{\rho_s}\right)^{1.65}$	$\left(\frac{\rho}{\rho_s}\right)^{1.8} < \frac{\lambda}{\lambda_s} < \left(\frac{\rho}{\rho_s}\right)^{1.65}$
Thermal exp. (10^{-6}/K), α	As solid	As solid
Latent heat (kJ/kg), L	As solid	As solid

(c) Scaling laws for electrical properties

Electrical properties	Open-cell foam	Closed-cell foam
Resistivity (10^{-8} ohm.m), R	$\left(\frac{\rho}{\rho_s}\right)^{-1.6} < \frac{R}{R_s} < \left(\frac{\rho}{\rho_s}\right)^{-1.85}$	$\left(\frac{\rho}{\rho_s}\right)^{-1.6} < \frac{R}{R_s} < \left(\frac{\rho}{\rho_s}\right)^{-1.85}$

whether a metal foam is a potential candidate. Examples of their use in this way are given in Chapters 10 and 11.

References

Banhart, J. (1999) Private communication.

CES (1999) *The Cambridge Engineering Selector*, Granta Design Ltd, Trumpington Mews, 40B High Street, Trumpington, Cambridge CB2 2LS, UK; tel: +44 1223 518895; fax: +44 1223 506432; web site: http//www.granta.co.uk

Degisher, P. (1999) Private communication.

Evans, A.G. (1997/1998/1999) (ed.) *Ultralight Metal Structures*, Division of Applied Sciences, Harvard University, Cambridge, Mass, USA: the annual report on the MURI programme sponsored by the Defence Advanced Research Projects Agency and Office of Naval Research.

Gibson, L.J. and Ashby, M.F. (1997) *Cellular Solids, Structure and Properties*, 2nd edition, Cambridge University Press, Cambridge.

Simancik, F. (1999) Private communication.

Chapter 5

Design analysis for material selection

A material is selected for a given component because its *property profile* matches that demanded by the application. The desired property profile characterizes the application. It is identified by examining the *function* of the component, the *objectives* which are foremost in the designer's mind, and *constraints* that the component must meet if it is to perform adequately.

The applications in which a new material will excel are those in which the match between its property profile and that of the application are particularly good. This chapter describes how property profiles are established. The Appendix to this Guide lists material-property groups linked with a range of generic applications. Metal foams have large values of some of these property groups and poor values of others, suggesting where applications might be sought.

5.1 Background

A property profile is a statement of the characteristics required of a material if it is to perform well in a given application. It has several parts. First, it identifies simple property limits which are dictated by constraints imposed by the design: requirements for electrical insulation or conduction impose limits on resistivity; requirements of operating temperature or environment impose limits on allowable service temperature or on corrosion and oxidation resistance. Second, it identifies material indices which capture design objectives: minimizing weight, perhaps, or minimizing cost, or maximizing energy storage. More precisely, a material index is a grouping of material properties which, if maximized or minimized, maximizes some aspect of the performance of an engineering component. Familiar indices are the specific stiffness, E/ρ, and the specific strength, σ_y/ρ, (where E is Young's modulus, σ_y is the yield strength or elastic limit, and ρ is the density), but there are many others. They guide the optimal selection of established materials and help identify potential applications for new materials. Details of the method, with numerous examples are given in Ashby (1999). PC-based software systems that implement the method are available (see, for example, *CES*, 1999).

Section 5.2 summarizes the way in which property profiles, indices and limits are derived. Section 5.3 gives examples of the method. Metal foams have particularly attractive values of certain indices and these are discussed in Section 5.4. A catalogue of indices can be found in the Appendix at the end of this Design Guide.

5.2 Formulating a property profile

The steps are as follows:

1. *Function*. Identify the *primary function* of the component for which a material is sought. A beam carries bending moments; a heat-exchanger tube transmits heat; a bus-bar transmits electric current.
2. *Objective*. Identify the *objective*. This is the first and most important quantity you wish to minimize or maximize. Commonly, it is weight or cost; but it could be energy absorbed per unit volume (a compact crash barrier); or heat transfer per unit weight (a light heat exchanger) – it depends on the application.
3. *Constraints*. Identify the *constraints*. These are performance goals that *must* be met, and which therefore limit the optimization process of step 2. Commonly these are: a required value for stiffness, S; for the load, F, or moment, M, or torque, T, or pressure, p, that must be safely supported; a given operating temperature implying a lower limit for the maximum use temperature, T_{max}, of the material; or a requirement that the component be electrically insulating, implying a limit on its resistivity, R.

   It is essential to distinguish between *objectives* and *constraints*. As an example, in the design of a racing bicycle, minimizing weight might be the objective with stiffness, toughness, strength and cost as constraints ('as light as possible without costing more than $500'). But in the design of a shopping bicycle, minimizing cost becomes the objective, and weight becomes a constraint ('as cheap as possible, without weighing more than 25 kg').
4. *Free variables*. The first constraint is one of geometry: the length, ℓ, and the width, b, of the panel are specified above but the thickness, t, is not – it is a *free variable*.
5. Lay out this information as in Table 5.1.
6. *Property limits*. The next three constraints impose simple *property limits*; these are met by choosing materials with adequate safe working temperature, which are electrical insulators and are non-magnetic.
7. *Material indices*. The final constraint on strength ('plastic yielding') is more complicated. Strength can be achieved in several ways: by choice of material, by choice of area of the cross-section, and by choice of cross-section shape (rib-stiffened or sandwich panels are examples), all of which

Table 5.1 *Design requirements*

Function	Panel to support electronic signal-generating equipment (thus carries bending moments)
Objective	Minimize weight
Constraints	Must have length, ℓ, and width, b,
	Must operate between −20°C and 120°C
	Must be electrically insulating
	Must be non-magnetic
	Must not fail by plastic yielding
Free variable	The thickness, t, of the panel

Table 5.2 *Deriving material indices: the recipe (Ashby, 1999)*

(a) Identify the aspect of *performance*, **P** (mass, cost, energy, etc.) to be maximized or minimized, defining the *objective* (mass, in the example of Table 5.1)
(b) Develop an *equation* for **P** (called the *objective function*)
(c) Identify the *free variables* in the objective function. These are the variables not specified by the design requirement (the thickness, t, of the panel in the example of Table 5.1)
(d) Identify the *constraints*. Identify those that are simple (independent of the variables in the objective function) and those which are dependent (they depend on the free variables in the objective function)
(e) Develop *equations* for the dependent constraints (no yield; no buckling, etc.)
(f) *Substitute* for the free variables from the equations for the dependent constraints into the objective function, eliminating the free variable(s)
(g) *Group the variables* into three groups: functional requirements, **F**, geometry, **G**, and material properties, **M**, thus:

$$\text{Performance } \mathbf{P} \leqslant \mathbf{f(F, G, M)}$$

(h) Read off the *material index*, **M**, to be maximized or minimized

impact the objective since they influence weight. This constraint must be coupled to the objective. To do this, we identify one or more *material indices* appropriate to the function, objective and constraints. The material index allows the optimization step of the selection. The method, in three stages, is as follows:

(i) Write down an equation for the *objective*
(ii) Eliminate the *free variable(s)* in this equation by using the *constraints*
(iii) Read off the grouping of material properties (called the *material index*) which maximize or minimize the objective.

A more detailed recipe is given in Table 5.2. Indices for numerous standard specifications are listed in the Appendix at the end of this Design Guide.

5.3 Two examples of single-objective optimization

Panel of specified stiffness and minimum mass

The mode of loading which most commonly dominates in engineering is not tension, but bending. Consider the performance metric for a panel of specified length, ℓ, and width, b (Figure 5.1), and specified stiffness, with the objective of minimizing its mass, m. The mass is

$$m = bt\rho \tag{5.1}$$

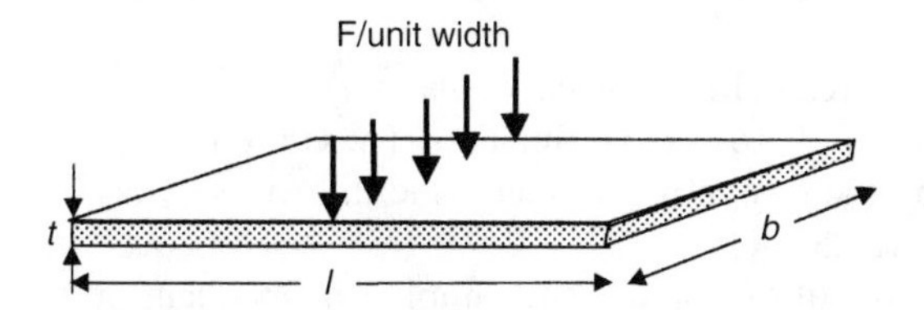

Figure 5.1 *A panel of length, ℓ, width, b, and thickness, t, loaded in bending by a force, F, per unit width*

where t is the thickness of the panel and ρ is the density of the material of which it is made. The length, ℓ, width, b, and force, F, per unit width are specified; the thickness, t, is free. We can reduce the mass by reducing t, but there is a lower limit set by the requirement that the panel must meet the constraint on its bending stiffness, S, meaning that it must not deflect more than δ under a load Fb. To achieve this we require that

$$S = \frac{Fb}{\delta} = \frac{B_1 EI}{\ell^3} \geq S^* \tag{5.2}$$

where S^* is the desired bending stiffness, E is Young's modulus, B_1 is a constant which depends on the distribution of load (tabulated in Chapter 6, Section 6.3) and I is the second moment of the area of the section. This, for a panel of section $b \times t$, is

$$I = \frac{bt^3}{12} \tag{5.3}$$

Using equations (5.2) and (5.3) to eliminate t in equation (5.1) gives the performance equation for the performance metric, m:

$$m \geq \left(\frac{12S^* b^2}{B_1}\right)^{1/3} \ell^2 \left(\frac{\rho}{E^{1/3}}\right) \quad (5.4)$$

This equation for the performance metric, m, is the objective function – it is the quantity we wish to minimize.

All the quantities in equation (5.4) are specified by the design except the group of material properties in the last bracket, $\rho/E^{1/3}$. This is the material index for the problem. The values of the performance metric for competing materials scale with this term. Taking material M_0 as the reference (the incumbent in an established design, or a convenient standard in a new one), the performance metric of a competing material M_1 differs from that of M_0 by the factor

$$\frac{m_1}{m_0} = \frac{(\rho_1/E_1^{1/3})}{(\rho_0/E_0^{1/3})} \quad (5.5)$$

where the subscript '0' refers to M_0 and the '1' to M_1.

Panel of specified strength and minimum mass

If, for the panel of Figure 5.1, the constraint were that of bending strength rather than stiffness, the constraining equation becomes that for failure load, F_f, per unit width, meaning the onset of yielding:

$$F_f = \frac{B_2 \sigma_y I}{b t \ell} \geq F_f^* \quad (5.6)$$

where B_2, like B_1, is a constant that depends only on the distribution of the load; it is tabulated in Chapter 6, Section 6.4. The performance metric, again, is the mass, m:

$$m \geqslant \left(\frac{6F_f^* b^2}{B_2}\right)^{1/2} \ell^{3/2} \left(\frac{\rho}{\sigma_y^{1/2}}\right) \quad (5.7)$$

where σ_y the yield strength of the material of which the panel is made and $F_f^* b$ is the desired minimum failure load. Here the material index is $\rho/\sigma_y^{1/2}$. Taking material M_0 as the reference again, the performance metric of a competing material M_1 differs from that of M_0 by the factor

$$\frac{m_1}{m_0} = \frac{(\rho_1/\sigma_{y,1}^{1/2})}{(\rho_0/\sigma_{y,0}^{1/2})} \quad (5.8)$$

More generally, if the performance metrics for a reference material M_0 are known, those for competing materials are found by scaling those of M_0 by the ratio of their material indices. There are many such indices. A few of those that appear most commonly are listed in Table 5.3. More are listed in the Appendix.

Table 5.3 *Material indices*[a]

Function, objective and constraint (and example)	*Index*
Tie, minimum weight, stiffness prescribed (cable support of a lightweight stiffness-limited tensile structure)	ρ/E
Tie, minimum weight, stiffness prescribed (cable support of a lightweight strength-limited tensile structure)	ρ/σ_y
Beam, minimum weight, stiffness prescribed (aircraft wing spar, golf club shaft)	$\rho/E^{1/2}$
Beam, minimum weight, strength prescribed (suspension arm of car)	$\rho/\sigma_y^{2/3}$
Panel, minimum weight, stiffness prescribed (car door panel)	$\rho/E^{1/3}$
Panel, minimum weight, strength prescribed (table top)	$\rho/\sigma_y^{1/2}$
Column, minimum weight, buckling load prescribed (push-rod of aircraft hydraulic system)	$\rho/E^{1/2}$
Spring, minimum weight for given energy storage (return springs in space applications)	$\rho E/\sigma_y^2$
Precision device, minimum distortion, temperature gradients prescribed (gyroscopes; hard-disk drives; precision measurement systems)	α/λ
Heat sinks, maximum thermal flux, thermal expansion prescribed (heat sinks for electronic systems)	α/λ
Electromagnet, maximum field, temperature rise and strength prescribed (ultra-high-field magnets; very high-speed electric motors)	$1/\kappa C_p\rho$

Note: ρ = density; E = Young's modulus; σ_y = elastic limit; λ = thermal conductivity; α = thermal expansion coefficient; κ = electrical conductivity; C_p = specific heat.

[a]The derivation of these and many other indices can be found in Ashby (1999).

5.4 Where might metal foams excel?

Material indices help identify applications in which a material might excel. Material-selection charts like those shown in Chapters 4, 11 and 12 allow the values of indices for metal foams to be established and compared with those of other engineering materials. The comparison reveals that metal foams have interesting values of the following indices:

1. The index $E^{1/3}/\rho$ which characterizes the bending-stiffness of lightweight panels (E is Young's modulus and ρ the density). A foam panel is lighter, for the same stiffness, than one of the same material which is solid. By using the foam as the core of a sandwich structure (Chapter 10) even greater weight saving is possible. Metal foam sandwiches are lighter than plywood panels of the same stiffness, and can tolerate higher temperatures. Their weight is comparable with that of waffle-stiffened aluminum panels but they have lower manufacturing cost.
2. The index $\sigma_y^{1/2}/\rho$ which characterizes the bending-strength of lightweight panels (σ_y is the elastic limit). A foam panel is stronger, for a given weight, than one of the same material which is solid. Strength limited foam-core sandwich panels and shells can offer weight savings over conventional stringer-stiffened structures (Chapters 7 and 10).
3. The exceptional energy-absorbing ability of metal foams is characterized by the index $\sigma_{pl}\varepsilon_D$ which measures the energy absorbed in crushing the material up to its 'densification' strain ε_D (σ_{pl} is the plateau stress). Metal foams absorb as much energy as tubes, and do so from *any* direction (Chapter 11).
4. The index $\eta E^{1/3}/\rho$ which measures the ability of a panel to damp flexural vibrations (η is the mechanical loss coefficient). High values of this index capture both high natural flexural vibration frequencies of metal foams (suppressing resonance in the acoustic range) and the ability of the material to dissipate energy internally.
5. The index $C_p\rho\lambda$ which characterizes the time-scale for penetration of a thermal front through an insulating layer of given thickness; it also characterizes the total thermal energy lost in the insulation of an oven or furnace in a thermal cycle (C_p is the specific heat and λ is the thermal conductivity). In both cases low values of the index are sought; foams offer these.

References

Ashby, M.F. (1999) *Materials Selection in Mechanical Design*, Butterworth-Heinemann, Oxford.

Ashby, M.F. and Cebon, D. (1997) *Case Studies in Materials Selection*, Granta Design Ltd, Trumpington Mews, 40B High Street, Trumpington, Cambridge CB2 2LS, UK; tel: +44 1223 518895; fax: +44 1223 506432; web site: http//www.granta.co.uk

CES (1999) Granta Design Ltd, Trumpington Mews, 40B High Street, Trumpington, Cambridge CB2 2LS, UK; tel: +44 1223 518895; fax: +44 1223 506432; web site: http//www.granta.co.uk

Chapter 6

Design formulae for simple structures

The formulae assembled here are useful for approximate analyses of the response of structures to load (more details can be found in Young, (1989). Each involves one or more material property. Results for solid metals appear under heading (a) in each section. The properties of metal foams differ greatly from those of solid metals. Comments on the consequences of these differences appear under heading (b) in each section.

6.1 Constitutive equations for mechanical response

(a) Isotropic solids

The behavior of a component when it is loaded depends on the mechanism by which it deforms. A beam loaded in bending may deflect elastically; it may yield plastically; it may deform by creep; and it may fracture in a brittle or in a ductile way. The equation which describes the material response is known as a constitutive equation, which differ for each mechanism. The constitutive equation contains one or more material properties: Young's modulus, E, and Poisson's ratio, ν, are the material properties which enter the constitutive equation for linear-elastic deformation; the elastic limit, σ_y, is the material property which enters the constitutive equation for plastic flow; the hardness, H, enters contact problems; the toughness J_{IC} enters that for brittle fracture. Information about these properties can be found in Chapter 2.

The common constitutive equations for mechanical deformation are listed in Table 6.1. In each case the equation for uniaxial loading by a tensile stress, σ, is given first; below it is the equation for multi-axial loading by principal stresses σ_1, σ_2 and σ_3, always chosen so that σ_1 is the most tensile and σ_3 the most compressive (or least tensile) stress. They are the basic equations which determine mechanical response.

(b) Metal foams

Metal foams are approximately linear-elastic at very small strains. In the linear-elastic region Hooke's law (top box, Table 6.1) applies. Because they change

Table 6.1 *Constitutive equations for mechanical response*

Isotropic solids: elastic deformation	
Uniaxial	$\varepsilon_1 = \dfrac{\sigma_1}{E}$
General	$\varepsilon_1 = \dfrac{\sigma_1}{E} - \dfrac{\nu}{E}(\sigma_2 + \sigma_3)$
Isotropic solids: plastic deformation	
Uniaxial	$\sigma_1 \geqslant \sigma_y$
General	$\sigma_1 - \sigma_3 = \sigma_y \quad (\sigma_1 > \sigma_2 > \sigma_3)$ (Tresca) $\sigma_e \geqslant \sigma_y$ (Von Mises) with $\sigma_e^2 = \frac{1}{2}[(\sigma_1 - \sigma_2)^2 + (\sigma_2 - \sigma_3)^2 + (\sigma_3 - \sigma_1)^2]$
Metal foams: elastic deformation	
Uniaxial General	As isotropic solids – though some foams are anisotripic
Metal foams: plastic deformation	
Uniaxial	$\sigma_1 \geqslant \sigma_y$
General	$\hat{\sigma} \geqslant \sigma_y$ with $\hat{\sigma}^2 = \dfrac{1}{(1 + (\alpha/3)^2)}[\sigma_e^2 + \alpha^2\sigma_m^2]$ and $\sigma_m = \frac{1}{3}(\sigma_1 + \sigma_2 + \sigma_3)$
Material properties	
E = Young's modulus ν = Poisson's ratio	σ_y = Yield strength α = Yield constant

volume when deformed plastically (unlike fully dense metals), a hydrostatic pressure influences yielding. A constitutive equation which describes their plastic response is listed in Table 6.1. It differs fundamentally from those for fully dense solids. Details are given in Chapter 7.

6.2 Moments of sections

A beam of uniform section, loaded in simple tension by a force, F, carries a stress $\sigma = F/A$ where A is the area of the section (see Figure 6.1). Its response is calculated from the appropriate constitutive equation. Here the important characteristic of the section is its area, A. For other modes of loading, higher moments of the area are involved. Those for various common sections are given below and are defined as follows.

The second moment of area I measures the resistance of the section to bending about a horizontal axis (shown as a broken line). It is

$$I = \int_{section} y^2 b(y)\,\mathrm{d}y$$

where y is measured vertically and $b(y)$ is the width of the section at y. The moment K measures the resistance of the section to twisting. It is equal to the polar moment of area J for circular sections, where

$$J = \int_{section} 2\pi r^3\,\mathrm{d}r$$

where r is measured radially from the centre of the circular section. For non-circular sections K is less than J. The section modulus $Z = I/y_m$ (where y_m is the normal distance from the neutral axis of bending to the outer surface of the beam) determines the surface stress σ generated by a given bending moment, M:

$$\sigma = \frac{M y_m}{I} = \frac{M}{Z}$$

Finally, the moment H, defined by

$$H = \int_{section} y b(y)\,\mathrm{d}y$$

measures the resistance of the beam to fully plastic bending. The fully plastic moment for a beam in bending is

$$M_p = H\sigma_y$$

Thin or slender shapes may buckle locally before they yield or fracture. It is this which sets a practical limit to the thinness of tube walls and webs (see Section 6.5).

SECTION	A (m^2)	I (m^4)	K (m^4)	I/y_m (m^3)	H (m^3)
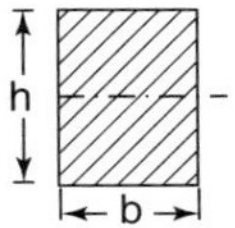 h, b	bh	$\frac{bh^3}{12}$	$\frac{16}{3}hb^3\left(1-0.58\frac{b}{h}\right)$	$\frac{bh^2}{6}$	$\frac{bh^2}{4}$
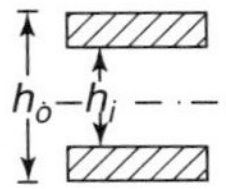 h_o, h_i	$b\,(h_o - h_i)$	$\frac{b}{12}(h_o^3 - h_i^3)$	—	$\frac{b}{12h_o}(h_o^3 - h_i^3)$	$\frac{b}{4}(h_o^2 - h_i^2)$
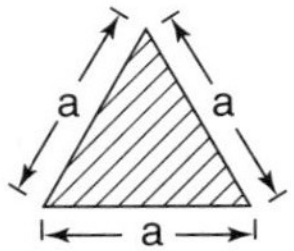 a, a, a	$\frac{\sqrt{3}}{4}a^2$	$\frac{a^4}{32\sqrt{3}}$	$\frac{a^4\sqrt{3}}{80}$	$\frac{a^3}{32}$	—
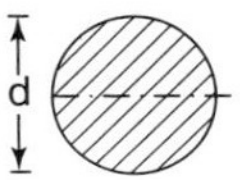 d	$\frac{\pi d^2}{4}$	$\frac{\pi}{64}d^4$	$\frac{\pi}{32}d^4$	$\frac{\pi}{32}d^3$	$\frac{1}{6}d^3$
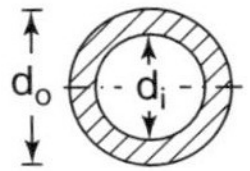 d_o, d_i	$\frac{\pi}{4}(d_o^2 - d_i^2)$	$\frac{\pi}{64}(d_o^4 - d_i^4)$	$\frac{\pi}{32}(d_o^4 - d_i^4)$	$\frac{\pi}{32d_o}(d_o^4 - d_i^4)$	$\frac{1}{6}(d_o^3 - d_i^3)$

Figure 6.1 *Moments of sections*

SECTION	A (m^2)	I (m^4)	K (m^4)	I/y_m (m^3)	H (m^3)
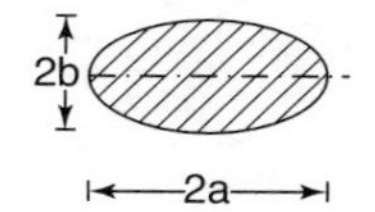 2b, 2a	πab	$\frac{\pi}{4} a b^3$	$\frac{\pi a^3 b^3}{a^2 + b^2}$	$\frac{\pi}{2} a b^2$	—
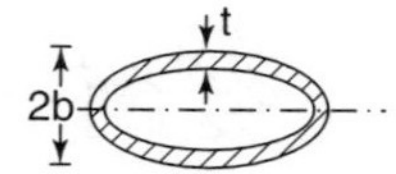 2b, t	$2\pi (ab)^{\frac{1}{2}} t$	$\frac{\pi}{4} a b^3 t \left(\frac{1}{a} + \frac{3}{b}\right)$	$\frac{4\pi t a^2 b^2}{(a+b)}$	$\frac{\pi a b^2 t}{2} \left(\frac{1}{a} + \frac{3}{b}\right)$	—
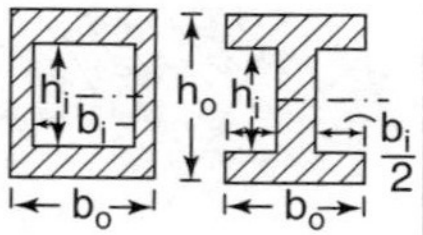 h_i, b_i, h_o, b_o, $\frac{b_i}{2}$	$h_o b_o - h_i b_i$	$\frac{1}{12}(b_o h_o^3 - b_i h_i^3)$	—	$\frac{1}{12 h_o}(b_o h_o^3 - b_i h_i^3)$	$\frac{1}{4}(b_o h_o^2 - b_i h_i^2)$
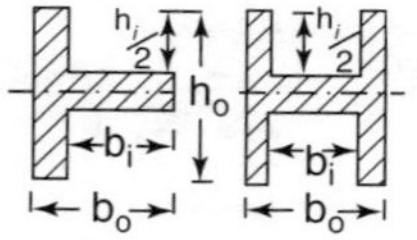 $\frac{h_i}{2}$, h_o, b_i, b_o	$h_o b_i + h_i b_o$	$\frac{1}{12}(b_i h_o^3 - b_o h_i^3)$	—	$\frac{1}{12 h_o}(b_i h_o^3 - b_o h_i^3)$	$\frac{1}{4}(b_o h_o^2 + b_i h_i^2 - 2b h_i h_o)$
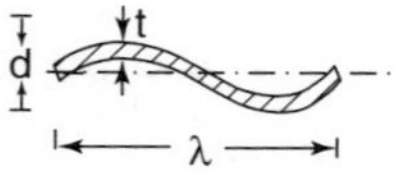 d, t, λ	$t\lambda \left(1 + \left(\frac{\pi d}{2\lambda}\right)^2\right)$	$\frac{t \lambda d^2}{8}\left(1 - \frac{0.81}{1 + 2.5\left(\frac{d}{2\lambda}\right)^2}\right)$	—	—	—

Figure 6.1 *(continued)*

6.3 Elastic deflection of beams and panels

(a) Isotropic solids

When a beam is loaded by a force, F, or moments, M, the initially straight axis is deformed into a curve. If the beam is uniform in section and properties, long in relation to its depth and nowhere stressed beyond the elastic limit, the deflection, δ, and the angle of rotation, θ, can be calculated from elastic beam theory. The differential equation describing the curvature of the beam at a point x along its length for small strains is

$$EI\frac{d^2y}{dx^2} = M(x)$$

where y is the lateral deflection, and $M(x)$ is the bending moment at the point x on the beam. E is Young's modulus and I is the second moment of area (Section 6.2). When M is constant, this becomes

$$\frac{M}{I} = E\left(\frac{1}{R} - \frac{1}{R_0}\right)$$

where R_0 is the radius of curvature before applying the moment and R the radius after it is applied. Deflections, δ, and rotations, θ, are found by integrating these equations along the beam. Equations for the deflection, δ, and end slope, θ, of beams, for various common modes of loading are shown below.

The stiffness of the beam is defined by

$$S = \frac{F}{\delta} = \frac{B_1 EI}{\ell^3}$$

It depends on Young's modulus, E, for the material of the beam, on its length, l, and on the second moment of its section, I. Values of B_1 are listed below.

(b) Metal foams

The moduli of open-cell metal foams scales as $(\rho/\rho_s)^2$, that of closed-cell foams has an additional linear term (Table 4.2). When seeking bending stiffness at low weight, the material index characterizing performance (see Appendix) is $E^{1/2}/\rho$ (beams) or $E^{1/3}/\rho$ (panels (see Figure 6.2)). Used as beams, foams have approximately the same index value as the material of which they are made; as panels, they have a higher one, meaning that the foam panel is potentially lighter for the same bending stiffness. Their performance, however, is best exploited as cores for sandwich structures (Chapter 10). Clamping metal foams requires special attention: (see Section 6.7).

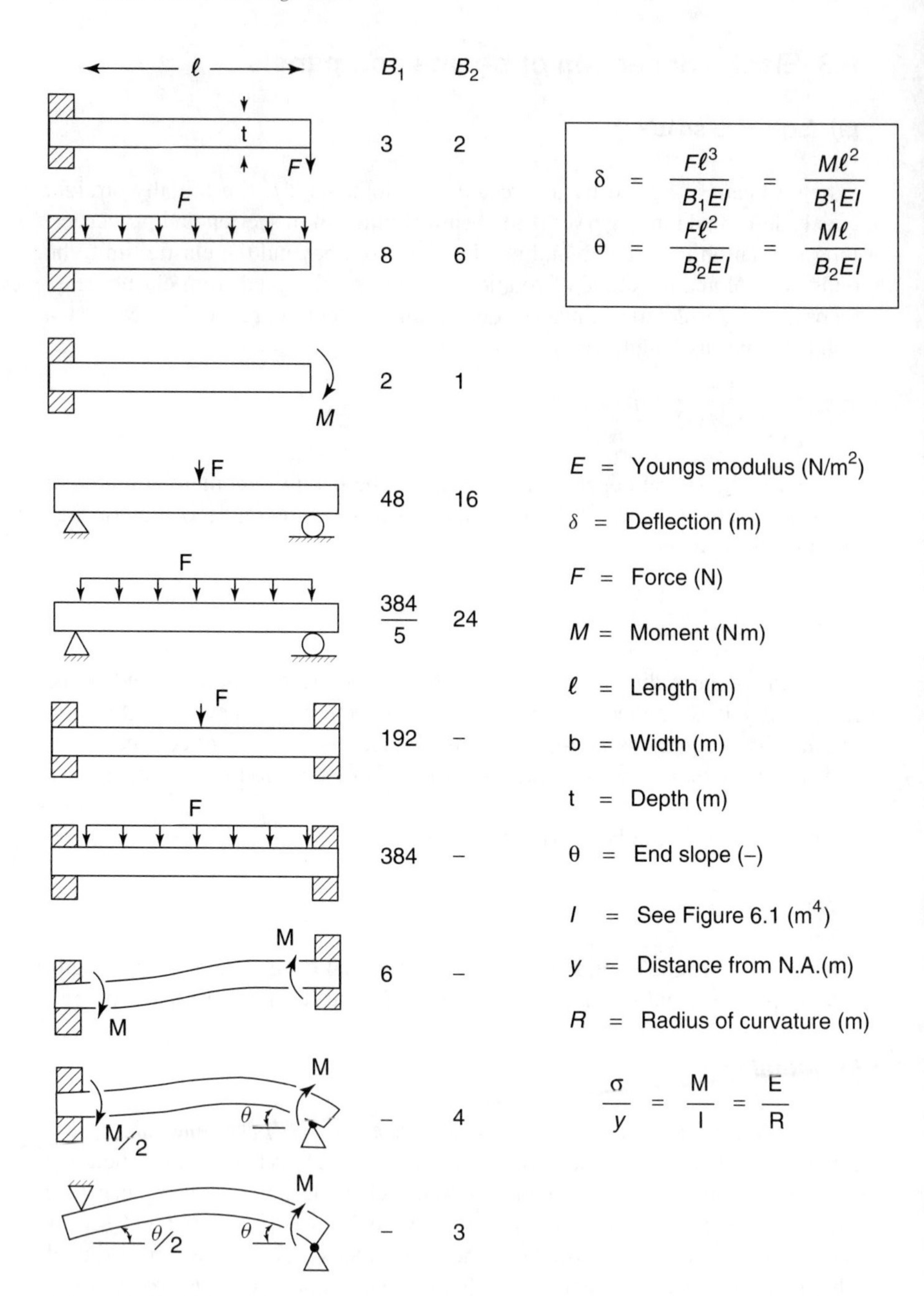

Figure 6.2 *Elastic bending of beams and panels*

6.4 Failure of beams and panels

(a) Isotropic solids

The longitudinal (or 'fiber') stress, σ, at a point, y, from the neutral axis of a uniform beam loaded elastically in bending by a moment, M, is

$$\frac{\sigma}{y} = \frac{M}{I} = E\left(\frac{1}{R} - \frac{1}{R_0}\right)$$

where I is the second moment of area (Section 6.2), E is Young's modulus, R_0 is the radius of curvature before applying the moment and R is the radius after it is applied. The tensile stress in the outer fiber of such a beam is

$$\sigma = \frac{M y_m}{I}$$

where y_m is the perpendicular distance from the neutral axis to the outer surface of the beam. If this stress reaches the yield strength, σ_y, of the material of the beam, small zones of plasticity appear at the surface (top diagram, Figure 6.3). The beam is no longer elastic, and, in this sense, has failed. If, instead, the maximum fiber stress reaches the brittle fracture strength, σ_f (the 'modulus of rupture', often shortened to MOR) of the material of the beam, a crack nucleates at the surface and propagates inwards (second diagram in Figure 6.3); in this case, the beam has certainly failed. A third criterion for failure is often important: that the plastic zones penetrate through the section of the beam, linking to form a plastic hinge (third diagram in Figure 6.3).

The failure moments and failure loads for each of these three types of failure and for each of several geometries of loading are given in Figure 6.3. The formulae labeled ONSET refer to the first two failure modes; those labeled FULL PLASTICITY refer to the third. Two new functions of section shape are involved. Onset of failure involves the quantity $Z = I/y_m$; full plasticity involves the quantity H (see Figure 6.3).

(b) Metal foams

The strength of open-cell metal foams scales as $(\rho/\rho_s)^{3/2}$, that of closed-cell foams has an additional linear term (Table 4.2). When seeking bending strength at low weight, the material index characterizing performance (see Appendix) is $\sigma_y^{3/2}/\rho$ (beams) or $\sigma_y^{1/2}/\rho$ (panels). Used as beams, foams have approximately the same index value as the material of which they are made; as panels, they have a higher one, meaning that, for a given bend strength, foam panels can be lighter. Clamping metal foams requires special attention: see Section 6.7.

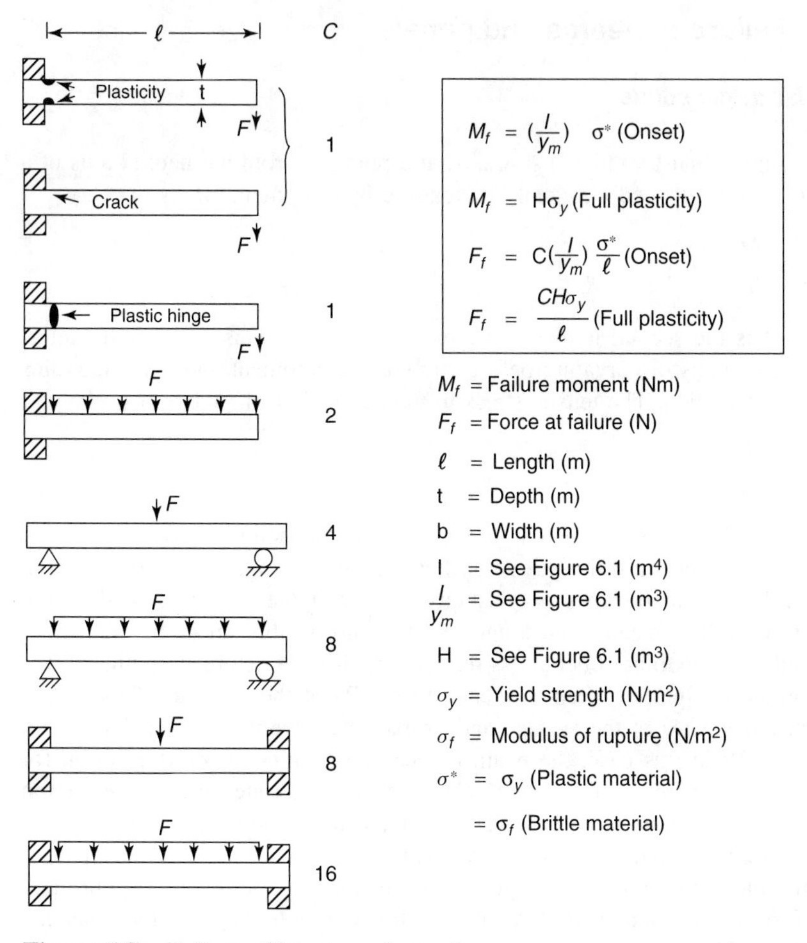

Figure 6.3 *Failure of beams and panels*

6.5 Buckling of columns, panels and shells

(a) Isotropic solids

If sufficiently slender, an elastic column, loaded in compression, fails by elastic buckling at a critical load, F_{crit}. This load is determined by the end constraints, of which four extreme cases are illustrated in Figure 6.4: an end may be constrained in a position and direction; it may be free to rotate but not translate (or 'sway'); it may sway without rotation; and it may both sway and rotate. Pairs of these constraints applied to the ends of column lead to the cases shown in the figure. Each is characterized by a value of the constant, n, which is equal to the number of half-wavelengths of the buckled shape.

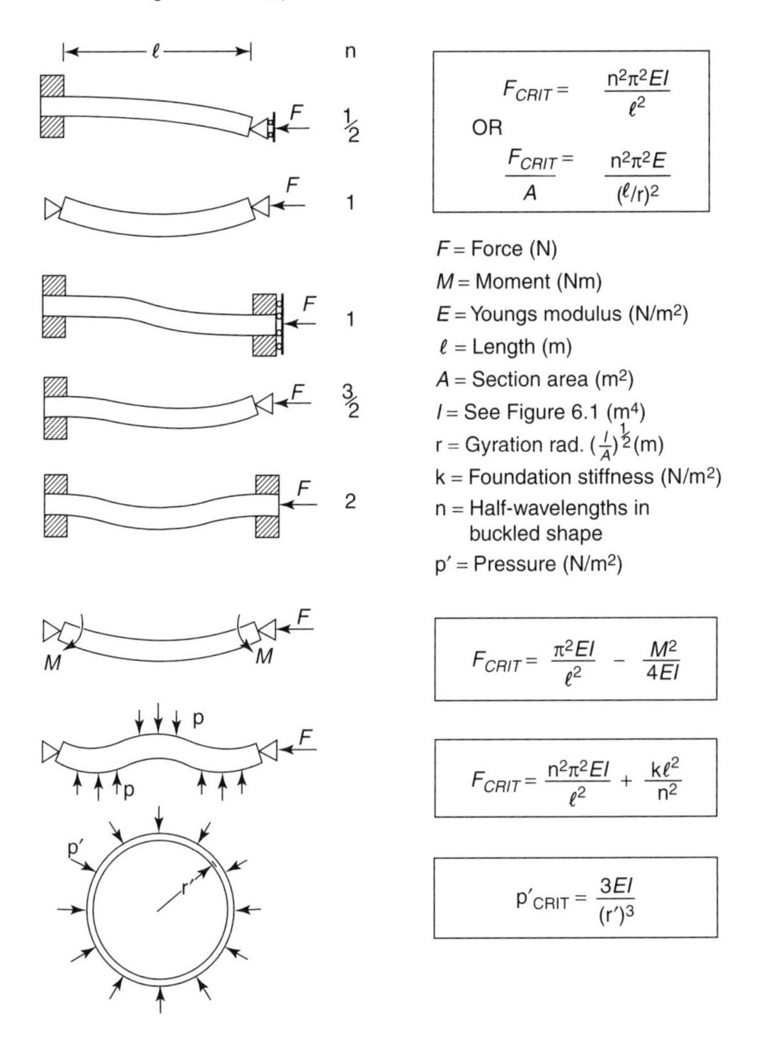

Figure 6.4 *Buckling of columns, panels and shells*

The addition of the bending moment, M, reduces the buckling load by the amount shown in the second box in Figure 6.4. A negative value of F_{crit} means that a tensile force is necessary to prevent buckling.

An elastic foundation is one that exerts a lateral restoring pressure, p, proportional to the deflection ($p = ky$ where k is the foundation stiffness per unit depth and y the local lateral deflection). Its effect is to increase F_{crit} by the amount shown in the third box.

A thin-walled elastic tube will buckle inwards under an external pressure p', given in the last box. Here I refers to the second moment of area of a section of the tube wall cut parallel to the tube axis.

(b) Metal foams

The moduli of open-cell metal foams scale as $(\rho/\rho_s)^2$, that of closed-cell foams has an additional linear term (Table 4.2). When seeking elastic-buckling resistance at low weight, the material index characterizing performance (see Appendix) is $E^{1/2}/\rho$ (beams) or $E^{1/3}/\rho$ (panels). As beam-columns, foams have the same index value as the material of which they are made; as panels, they have a higher one, meaning that the foam panel is potentially lighter for the same buckling resistance. Sandwich structures with foam cores (Chapter 10) are better still. Clamping metal foams requires special attention: see Section 6.7.

6.6 Torsion of shafts

(a) Isotropic solids

A torque, T, applied to the ends of an isotropic bar of uniform section, and acting in the plane normal to the axis of the bar, produces an angle of twist θ. The twist is related to the torque by the first equation below, in which G is the shear modulus. For round bars and tubes of circular section, the factor K is equal to J, the polar moment of inertia of the section, defined in Section 6.2. For any other section shape K is less than J. Values of K are given in Section 6.2.

If the bar ceases to deform elastically, it is said to have failed. This will happen if the maximum surface stress exceeds either the yield strength, σ_y, of the material or the stress at which it fractures. For circular sections, the shear stress at any point a distance r from the axis of rotation is

$$\tau = \frac{Tr}{K} = \frac{G\theta_r}{\ell}$$

The maximum shear stress, τ_{max}, and the maximum tensile stress, σ_{max}, are at the surface and have the values

$$\tau_{max} = \sigma_{max} = \frac{Td_0}{2K} = \frac{G\theta d_0}{2\ell}$$

If τ_{max} exceeds $\sigma_y/2$ (using a Tresca yield criterion), or if σ_{max} exceeds the MOR, the bar fails, as shown in Figure 6.5. The maximum surface stress for the solid ellipsoidal, square, rectangular and triangular sections is at the

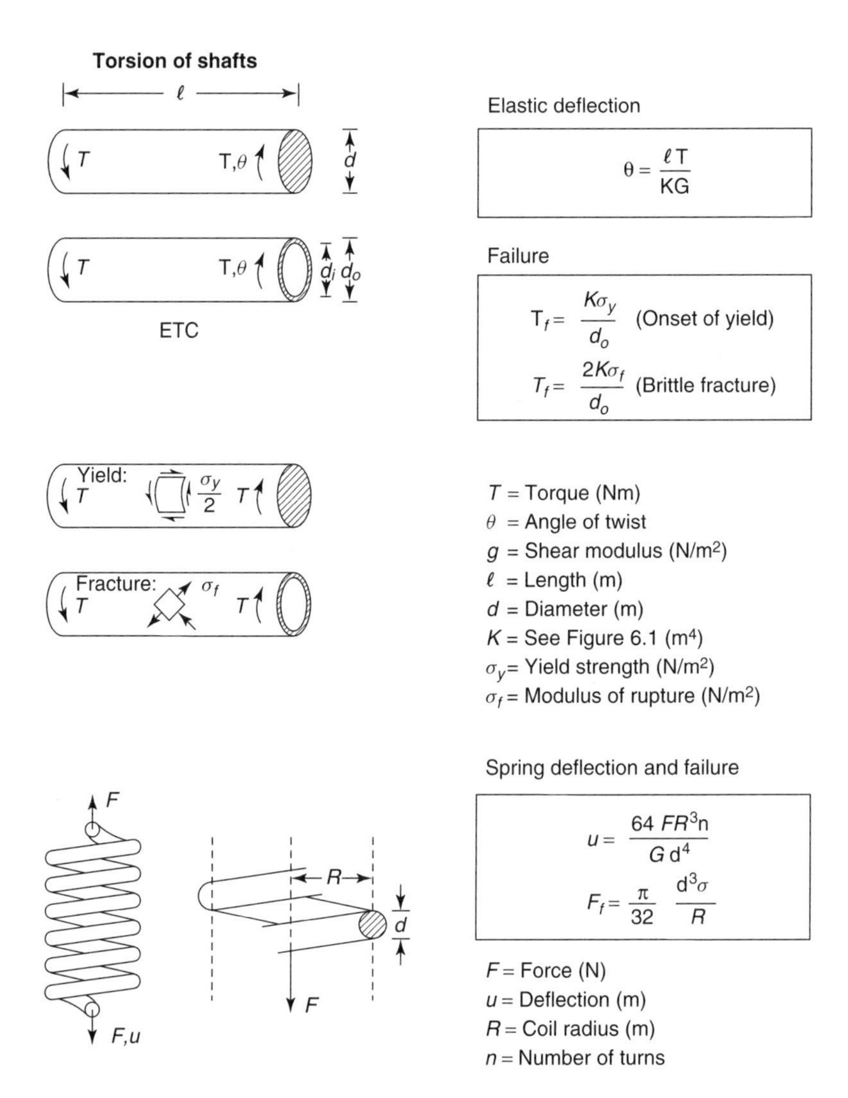

Figure 6.5 *Torsion of shafts*

points on the surface closest to the centroid of the section (the mid-points of the longer sides). It can be estimated approximately by inscribing the largest circle which can be contained within the section and calculating the surface stress for a circular bar of that diameter. More complex section-shapes require special consideration, and, if thin, may additionally fail by buckling. Helical springs are a special case of torsional deformation. The extension of a helical spring of n turns of radius R, under a force F, and the failure force F_{crit}, is given in Figure 6.5.

(b) Metal foams

The shear moduli of open-cell foams scales as $(\rho/\rho_s)^2$ and that of closed-cell foams has an additional linear term (Table 4.2). When seeking torsional stiffness at low weight, the material index characterizing performance (see Appendix) is G/ρ or $G^{1/2}/\rho$ (solid and hollow shafts). Used as shafts, foams have, at best, the same index value as the material of which they are made; usually it is less. Nothing is gained by using foams as torsion members.

6.7 Contact stresses

(a) Isotropic solids

When surfaces are placed in contact they touch at one or a few discrete points. If the surfaces are loaded, the contacts flatten elastically and the contact areas grow until failure of some sort occurs: failure by crushing (caused by the compressive stress, σ_c), tensile fracture (caused by the tensile stress, σ_t) or yielding (caused by the shear stress σ_s). The boxes in Figure 6.6 summarize the important results for the radius, a, of the contact zone, the centre-to-centre displacement u and the peak values of σ_c, σ_t and σ_s.

The first box in the figure shows results for a sphere on a flat, when both have the same moduli and Poisson's ratio has the value $\frac{1}{4}$. Results for the more general problem (the 'Hertzian Indentation' problem) are shown in the second box: two elastic spheres (radii R_1 and R_2, moduli and Poisson's ratios E_1, ν_1 and E_2, ν_2) are pressed together by a force F.

If the shear stress σ_s exceeds the shear yield strength $\sigma_y/2$, a plastic zone appears beneath the centre of the contact at a depth of about $a/2$ and spreads to form the fully plastic field shown in the second figure from the bottom of Figure 6.6. When this state is reached, the contact pressure (the 'indentation hardness') is approximately three times the yield stress, as shown in the bottom box:

$$H \approx 3\sigma_y$$

(b) Metal foams

Foams densify when compressed. The plastic constraint associated with indentation of dense solids is lost, and the distribution of displacements beneath the indent changes (bottom figure in Figure 6.6). The consequence: the indentation hardness of low-density foams is approximately equal to its compressive yield strength σ_c:

$$H \approx \sigma_c$$

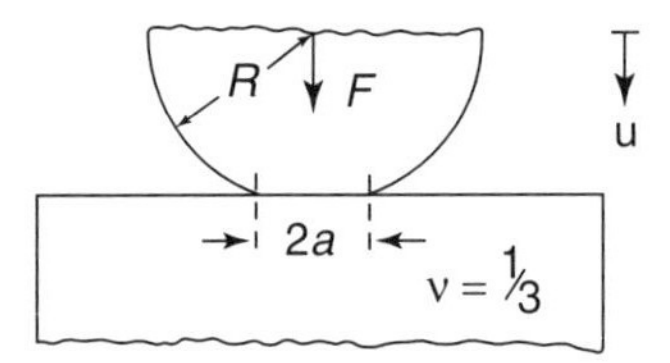

$$a = 0.7\left(\frac{FR}{E}\right)^{1/3}, \quad u = 1.0\left(\frac{F^2}{E^2 R}\right)^{1/3} \qquad \nu = \frac{1}{3}$$

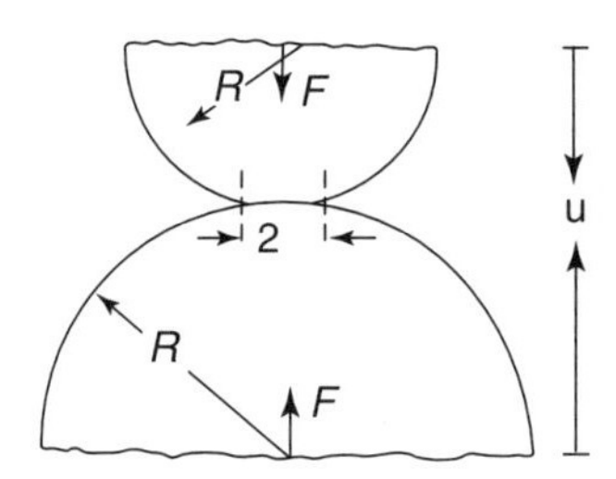

$$a = \left(\frac{3}{4}\frac{F}{E^*}\frac{R_1 R_2}{(R_1 + R_2)}\right)^{1/3}$$

$$u = \left(\frac{9}{16}\frac{F^2}{(E^*)^2}\frac{(R_1 + R_2)}{R_1 R_2}\right)^{1/3}$$

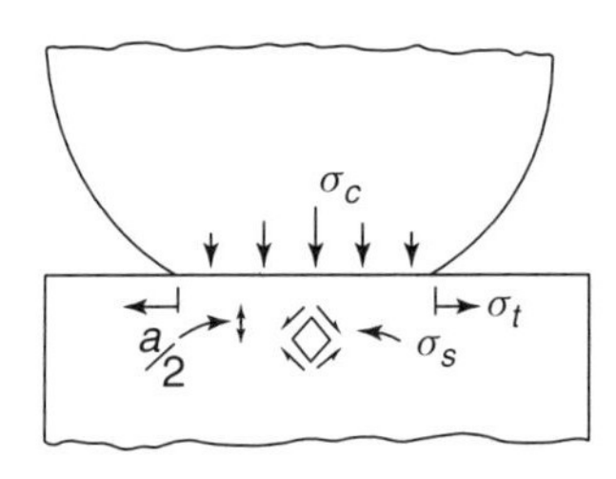

$$(\sigma_c)_{max} = \frac{3F}{2\pi a^2}$$

$$(\sigma_s)_{max} = \frac{F}{2\pi a^2}$$

$$(\sigma_t)_{max} = \frac{F}{6\pi a^2}$$

Plastic indent: dense solid

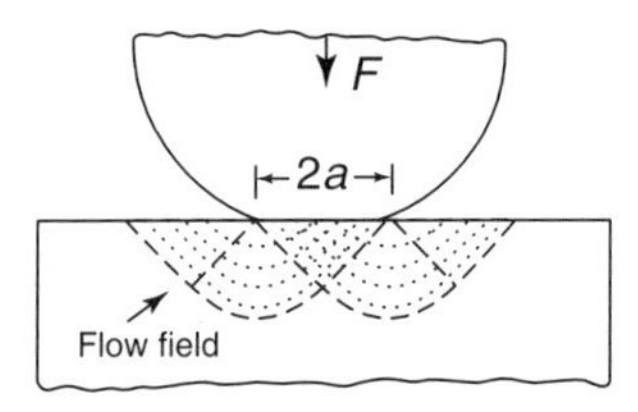

Plastic indent: metal foam

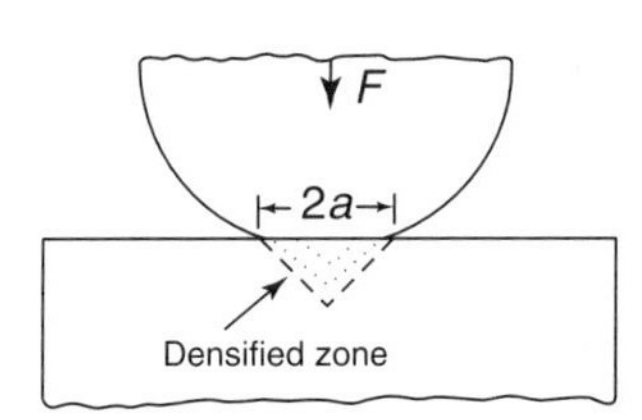

R_1 R_2	Radii of spheres (m)
E_1 E_2	Modulii of spheres (N/m^2)
ν_1 ν_2	Poission's ratios
F	Load (N)
a	Radius of contact (m)
u	Dadius of contact (m)
σ	Stresses (N/m^2)
σ_y	Yield stress (N/m^2)
E^*	$\left(\frac{1-\nu_1^2}{E_1} + \frac{1-\nu_2^2}{E_2}\right)^{-1}$

$$\frac{F}{\pi a^2} = 3\sigma_y$$

Figure 6.6 *Contact stress*

If the foam is not of low density, the indentation hardness is better approximated by

$$H \approx \sigma_c \left(1 + 2\left(\frac{\rho}{\rho_s}\right)\right)$$

This means that foams are more vulnerable to contact loads than dense solids, and that care must be taken in when clamping metal foams or joining them to other structural members: a clamping pressure exceeding σ_c will cause damage.

6.8 Vibrating beams, tubes and disks

(a) Isotropic solids

Anything vibrating at its natural frequency without significant damping can be reduced to the simple problem of a mass, m, attached to a spring of stiffness, K. The lowest natural frequency of such a system is

$$f = \frac{1}{2\pi}\sqrt{\frac{K}{m}}$$

Specific cases require specific values for m and K. They can often be estimated with sufficient accuracy to be useful in approximate modeling. Higher natural vibration frequencies are simple multiples of the lowest.

The first box in Figure 6.7 gives the lowest natural frequencies of the flexural modes of uniform beams with various end-constraints. As an example, the first can be estimated by assuming that the effective mass of the vibrating beam is one quarter of its real mass, so that

$$m = \frac{m_0 \ell}{4}$$

where m_0 is the mass per unit length of the beam (i.e. m is half the total mass of the beam) and K is the bending stiffness (given by F/δ from Section 6.3); the estimate differs from the exact value by 2%. Vibrations of a tube have a similar form, using I and m_0 for the tube. Circumferential vibrations can be found approximately by 'unwrapping' the tube and treating it as a vibrating plate, simply supported at two of its four edges.

The second box gives the lowest natural frequencies for flat circular disks with simply supported and clamped edges. Disks with curved faces are stiffer and have higher natural frequencies.

(b) Metal foams: scaling laws for frequency

Both longitudinal and flexural vibration frequencies are proportional to $\sqrt{E/\rho}$, where E is Young's modulus and ρ is the density, provided the dimensions of

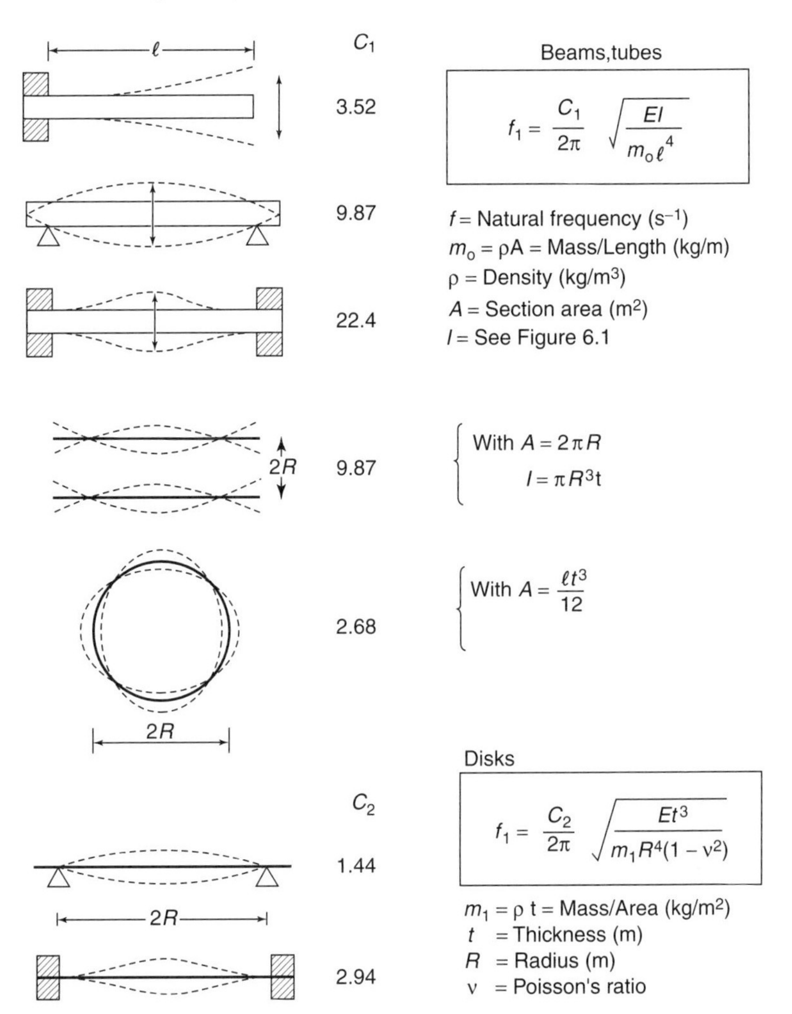

Figure 6.7 *Vibrating beams, tubes and disks*

the sample are fixed. The moduli of foams scale as $(\rho/\rho_s)^2$, and the mass as (ρ/ρ_s). Thus the natural vibration frequencies of a sample of fixed dimensions scale as $f/f_s = (\rho/\rho_s)^{1/2}$ – the lower the density of the foam, the lower its natural vibration frequencies. By contrast, the natural vibration frequencies of panels of the same stiffness (but of different thickness) scale as $f/f_s = (\rho/\rho_s)^{-1/6}$ – the lower the density, the higher the frequency. And for panels of equal *mass* (but of different thickness) the frequencies scale as $f/f_s = (\rho/\rho_s)^{-1/2}$ – the lower the density, the higher the frequency.

6.9 Creep

(a) Isotropic solids

Materials creep when loaded at temperatures above $1/3T_m$ (where T_m is the absolute melting point). It is convenient to characterize the creep of a material by its behavior under a tensile stress σ, at a temperature T_m. Under these conditions the tensile strain-rate $\dot{\varepsilon}$ is often found to vary as a power of the stress and exponentially with temperature:

$$\dot{\varepsilon} = A\left(\frac{\sigma}{\sigma_0}\right)^n \exp -\frac{Q}{RT}$$

where Q is an activation energy and R the gas constant. At constant temperature this becomes

$$\dot{\varepsilon} = \dot{\varepsilon}_0\left(\frac{\sigma}{\sigma_0}\right)^n$$

where $\dot{\varepsilon}_0(\text{s}^{-1})$, $\sigma_0(\text{N/m}^2)$ and n are creep constants.

The behavior of creeping components is summarized in Figure 6.8 which gives the deflection rate of a beam, the displacement rate of an indenter and the change in relative density of cylindrical and spherical pressure vessels in terms of the tensile creep constants.

(b) Metal foams

When foams are loaded in tension or compression the cell edges bend. When this dominates (as it usually does) the creep rate can be derived from the equation in the second box in Figure 6.8, with appropriate allowance for cell-edge geometry (see Chapter 9 for details). The resulting axial strain rate is given in the bottom box. The analogy between this and the equation in the top box suggests that the creep behavior of beams, plates tubes (and other structures) made of foam can be found from standard solutions for dense solids by replacing σ_0 by

$$\sigma_0^* = \left(\frac{n+2}{0.6}\right)^{1/n}\left\{\frac{n}{1.7(2n+1)}\right\}\left(\frac{\rho}{\rho_s}\right)^{(3n+1/2n)}\sigma_0$$

which, for large $n (n > 3)$, is well approximated by

$$\sigma_0^* \approx \frac{1}{2}\left(\frac{\rho}{\rho_s}\right)^{3/2}\sigma_0.$$

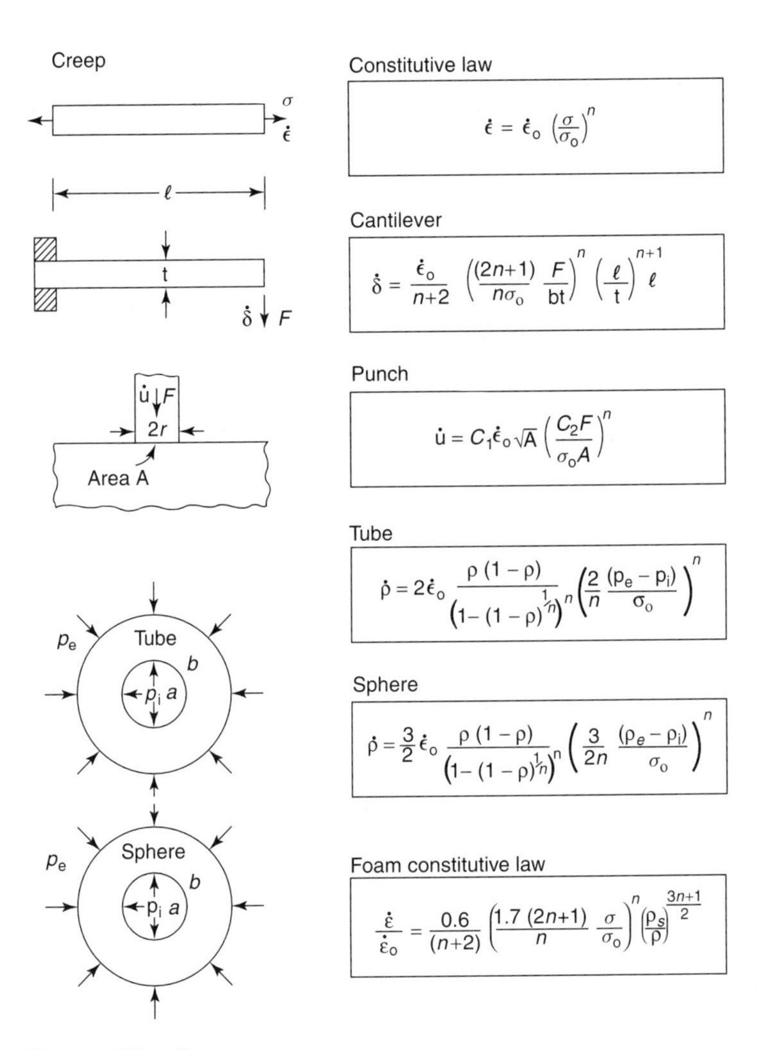

Figure 6.8 *Creep*

This is correct for simple tension and compression, and a reasonable approximation for bending and torsion, but it breaks down for indentation and hydrostatic compression because volumetric creep-compression of the foam has been neglected.

Reference

Young, W.C. (1989) *Roark's Formulas for Stress and Strain*, 6th edition, McGraw-Hill, New York.

Chapter 7

A constitutive model for metal foams

The plastic response of metal foams differs fundamentally from that of fully dense metals because foams compact when compressed, and the yield criterion is dependent on pressure. A constitutive relation characterizing this plastic response is an essential input for design with foams.

Insufficient data are available at present to be completely confident with the formulation given below, but it is consistent both with a growing body of experimental results, and with the traditional development of plasticity theory. It is based on recent studies by Deshpande and Fleck (1999, 2000), Miller (1999) and Gibson and co-workers (Gioux *et al.*, 1999, and Gibson and Ashby, 1997).

7.1 Review of yield behavior of fully dense metals

Fully dense metals deform plastically at constant volume. Because of this, the yield criterion which characterizes their plastic behavior is independent of mean stress. If the metal is isotropic (i.e. has the same properties in all directions) its plastic response is well approximated by the *von Mises criterion*: yield occurs when the von Mises effective stress σ_e attains the yield value Y.

The effective stress, σ_e, is a scalar measure of the deviatoric stress, and is defined such that it equals the uniaxial stress in a tension or compression test. On writing the principal stresses as $(\sigma_{\mathrm{I}}, \sigma_{\mathrm{II}}, \sigma_{\mathrm{III}})$, σ_e can be expressed by

$$2\sigma_e^2 = (\sigma_{\mathrm{I}} - \sigma_{\mathrm{II}})^2 + (\sigma_{\mathrm{II}} - \sigma_{\mathrm{III}})^2 + (\sigma_{\mathrm{III}} - \sigma_{\mathrm{I}})^2 \tag{7.1}$$

When the stress $\boldsymbol{\sigma}$ is resolved onto arbitrary Cartesian axes X_i, not aligned with the principal axes of stress, $\boldsymbol{\sigma}$ has three direct components $(\sigma_{11}, \sigma_{22}, \sigma_{33})$ and three shear components $(\sigma_{12}, \sigma_{23}, \sigma_{31})$, and can be written as a symmetric 3×3 matrix, with components σ_{ij}. Then, the mean stress σ_m, which is invariant with respect to a rotation of axes, is defined by

$$\sigma_m \equiv \tfrac{1}{3}(\sigma_{11} + \sigma_{22} + \sigma_{33}) = \tfrac{1}{3}\sigma_{kk} \tag{7.2}$$

where the repeated suffix, here and elsewhere, denotes summation from 1 to 3. The stress $\boldsymbol{\sigma}$ can be decomposed additively into its mean component σ_m

and its deviatoric (i.e. shear) components S_{ij}, giving

$$\sigma_{ij} = S_{ij} + \sigma_m \delta_{ij} \tag{7.3}$$

where δ_{ij} is the Kronecker delta symbol, and takes the value $\delta_{ij} = 1$ if $i = j$, and $\delta_{ij} = 0$ otherwise. The von Mises effective stress, σ_e, then becomes

$$\sigma_e^2 = \tfrac{3}{2} S_{ij} S_{ij} \tag{7.4}$$

In similar fashion, the strain rate $\dot{\varepsilon}$ is a symmetric 3×3 tensor, with three direct components ($\dot{\varepsilon}_{11}$, $\dot{\varepsilon}_{22}$, $\dot{\varepsilon}_{33}$) and three shear components ($\dot{\varepsilon}_{12}$, $\dot{\varepsilon}_{23}$, $\dot{\varepsilon}_{31}$). The volumetric strain rate is defined by

$$\dot{\varepsilon}_m \equiv \dot{\varepsilon}_{11} + \dot{\varepsilon}_{22} + \dot{\varepsilon}_{33} = \dot{\varepsilon}_{kk} \tag{7.5}$$

and the strain rate can be decomposed into its volumetric part $\dot{\varepsilon}_m$ and deviatoric part $\dot{\varepsilon}'_{ij}$ according to

$$\dot{\varepsilon}_{ij} = \dot{\varepsilon}'_{ij} + \tfrac{1}{3} \delta_{ij} \dot{\varepsilon}_m \tag{7.6a}$$

The strain rate $\dot{\varepsilon}$ can be written as the sum of an elastic strain rate $\dot{\varepsilon}^E$ and a plastic strain rate $\dot{\varepsilon}^P$. In an analogous manner to equation (7.6a), the plastic strain rate can be decomposed into an deviatoric rate $\dot{\varepsilon}^{P'}$ and a mean rate $\dot{\varepsilon}_m^P \equiv \dot{\varepsilon}_{kk}^P$, such that

$$\dot{\varepsilon}_{ij}^P = \dot{\varepsilon}_{ij}^{P'} + \tfrac{1}{3} \delta_{ij} \dot{\varepsilon}_m^P \tag{7.6b}$$

Now, for fully dense metallic solids, plastic flow occurs by slip with no change of volume, and so the volumetric plastic strain rate $\dot{\varepsilon}_m^P \equiv \dot{\varepsilon}_{kk}^P$ equals zero. Then, a useful scalar measure of the degree of plastic straining is the effective strain rate $\dot{\varepsilon}_e$, defined by

$$\dot{\varepsilon}_e^2 \equiv \tfrac{2}{3} \dot{\varepsilon}_{ij}^{P'} \dot{\varepsilon}_{ij}^{P'} \tag{7.7}$$

where the factor of $\frac{2}{3}$ has been introduced so that $\dot{\varepsilon}_e$ equals the uniaxial plastic strain rate in a tension (or compression) test on an incompressible solid.

In conventional Prandtl–Reuss J2 flow theory, the yield criterion is written

$$\Phi \equiv \sigma_e - Y \leqslant 0 \tag{7.8}$$

and the plastic strain rate $\dot{\varepsilon}_{ij}^P$ is normal to the yield surface Φ in stress space, and is given by

$$\dot{\varepsilon}_{ij}^P = \dot{\varepsilon}_e \frac{\partial \Phi}{\partial \sigma_{ij}} \tag{7.9}$$

(This prescription for the plastic strain rate enforces it to be incompressible, that is, $\dot{\varepsilon}_m^P = 0$.) The hardening rate is specified upon assuming that the effective strain rate $\dot{\varepsilon}_e$ scales with the effective stress rate $\dot{\sigma}_e$ according to

$$\dot{\varepsilon}_e \equiv \dot{\sigma}_e / h \tag{7.10}$$

where the hardening modulus, h, is the slope of the uniaxial stress versus plastic strain curve at a uniaxial stress of level σ_e. In all the above, true measures of stress and strain are assumed.

7.2 Yield behavior of metallic foams

We can modify the above theory in a straightforward manner to account for the effect of porosity on the yield criterion and strain-hardening law for a metallic foam. We shall assume the elastic response of the foam is given by that of an isotropic solid, with Young's modulus E and Poission's ratio ν. Since foams can yield under hydrostatic loading in addition to deviatoric loading, we modify the yield criterion (7.8) to

$$\Phi \equiv \hat{\sigma} - Y \leqslant 0 \tag{7.11}$$

where we define the *equivalent stress* $\hat{\sigma}$ by

$$\hat{\sigma}^2 \equiv \frac{1}{(1 + (\alpha/3)^2)} [\sigma_e^2 + \alpha^2 \sigma_m^2] \tag{7.12}$$

This definition produces a yield surface of elliptical shape in $(\sigma_m - \sigma_e)$ space, with a uniaxial yield strength (in tension and in compression) of Y, and a hydrostatic strength of

$$|\sigma_m| = \frac{\sqrt{(1 + (\alpha/3)^2}}{\alpha} Y$$

The parameter α defines the aspect ratio of the ellipse: in the limit $\alpha = 0$, $\hat{\sigma}$ reduces to σ_e and a J2 flow theory solid is recovered. Two material properties are now involved instead of one: the uniaxial yield strength, Y, and the pressure-sensitivity coefficient, α. The property Y is measured by a simple compression test, which can also be used to measure α in the way described below.

The yield surfaces for Alporas and Duocel for compressive stress states are shown in Figure 7.1. The data have been normalized by the uniaxial compressive yield strength, so that $\sigma_e = 1$ and $\sigma_m = \frac{1}{3}$ for the case of uniaxial compression. We note that the aspect ratio α of the ellipse lies in the range 1.35 to 2.08. The effect of yield surface shape is reflected in the measured

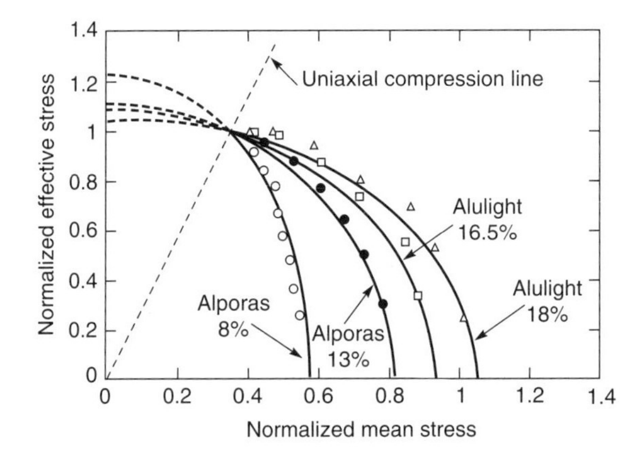

Figure 7.1 *Yield surfaces for Alporas and Duocel foams. The surfaces are approximately elliptical, described by equations (7.11) and (7.12)*

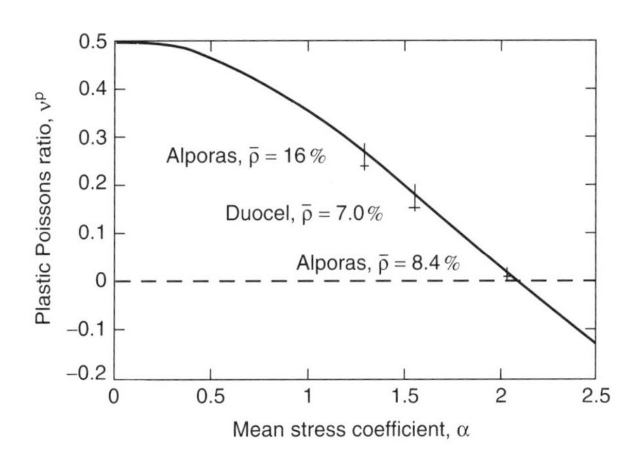

Figure 7.2 *The relationship between the plastic Poisson's ratio ν^P and the constant α*

plastic Poisson's ratio in a uniaxial compression test: the ratio of transverse strain to axial strain ν^P depends upon α, as shown in Figure 7.2. Experimental data, available for Alporas and Duocel foams, support this, (see Figure 7.2). The yield surface shape (equations (7.11) and (7.12)) is sufficiently simple for an analytical expression to be derivable for ν^P in terms of α, giving

$$\nu^P = \frac{\dfrac{1}{2} - \left(\dfrac{\alpha}{3}\right)^2}{1 + \left(\dfrac{\alpha}{3}\right)^2} \tag{7.13}$$

with the inversion

$$\alpha = 3\left(\frac{\frac{1}{2} - \nu^P}{1 + \nu^P}\right)^{1/2} \tag{7.14}$$

It would appear that the measurement of ν^P in a uniaxial compression test offers a quick and simple method for estimation of the value for α, and thereby the shape of the yield surface. Preliminary experience suggests that the measurement of ν^P is best done by compressing a sample, with suitably lubricated loading platens, to a uniaxial strain of 20–30%.

Having defined the yield surface shape, it remains to stipulate how the yield surface evolves with strain. For simplicity, we shall assume that isotropic hardening occurs: the yield surface grows in a geometrically self-similar manner with strain; the limited measurements of the yield surface for metallic foams approximate this behavior (see, for example, Figure 7.3 for the case of Alporas with an initial relative density of 0.16). Yield surfaces are displayed for the initial state, and for 10% and 30% uniaxial pre-strain. We note that the yield surfaces are smooth and geometrically self-similar.

We assume that the strain-hardening rate scales with the uniaxial compression response as follows. The plastic strain rate is again taken to be normal to the yield surface (7.11), and specified by the analogue of (7.9), given by

$$\dot{\varepsilon}_{ij}^P = \dot{\hat{\varepsilon}} \frac{\partial \Phi}{\partial \sigma_{ij}} \tag{7.15}$$

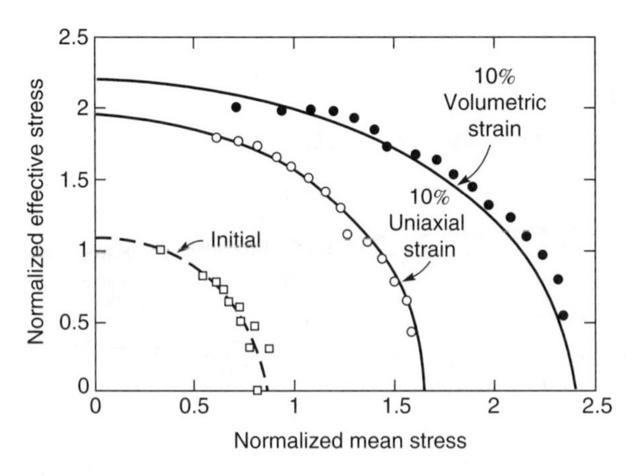

Figure 7.3 *The evolution of the yield surface with strain for an Alporas foam with an initial relative density of 0.16*

where the equivalent strain rate $\dot{\hat{\varepsilon}}$ is the work conjugate of $\hat{\sigma}$, such that

$$\hat{\sigma}\dot{\hat{\varepsilon}} = \sigma_{ij}\dot{\varepsilon}_{ij}^{P} \tag{7.16a}$$

and can be written explicitly as

$$\dot{\hat{\varepsilon}}^2 \equiv (1 + (\alpha/3)^2)\left[\dot{\varepsilon}_e^2 + \frac{1}{\alpha^2}\dot{\varepsilon}_m^{P^2}\right] \tag{7.16b}$$

A unique hardening relation is given by the evolution of $\hat{\sigma}$ with $\hat{\varepsilon}$. Unless otherwise stated, the uniaxial compressive stress–strain response is used to define the $\hat{\sigma} - \hat{\varepsilon}$ relation, as follows. The true stress σ versus logarithmic plastic strain ε^P curve in uniaxial compression is written in the incremental form

$$\dot{\varepsilon}^P = \dot{\sigma}/h(\sigma) \tag{7.17}$$

where the slope h evolves with increasing stress level σ. Recall that, for the case of uniaxial compression (or tension), the above definitions of $\hat{\sigma}$ and of $\dot{\hat{\varepsilon}}$ have been so normalized that $\hat{\sigma}$ is the uniaxial stress and $\dot{\hat{\varepsilon}}$ is the uniaxial plastic strain rate. The hardening law (7.17) for uniaxial loading can then be rewritten as

$$\dot{\hat{\varepsilon}} = \dot{\hat{\sigma}}/h(\hat{\sigma}) \tag{7.18}$$

It is assumed that this relation holds also for general multi-axial loading. Some checks on the accuracy of this approach are given in Figure 7.4: the measured tensile, compressive and shear stress–strain curves for Alporas and Al 6101-T6 Duocel foams are shown in terms of $\hat{\sigma}$ versus $\hat{\varepsilon}$. It is noted that

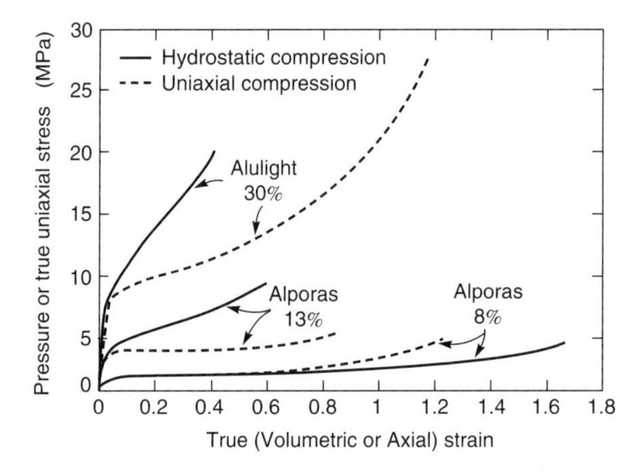

Figure 7.4 *Demonstration of the ability of the equivalent stress $\hat{\sigma}$ and the equivalent strain $\hat{\varepsilon}$ to define uniquely the stress–strain response of Alporas ($\overline{\rho} = 0.16$) and Duocel ($\overline{\rho} = 0.071$) foams. The tension, compression and shear response are plotted in terms of $\hat{\sigma}$ and $\hat{\varepsilon}$*

the curves almost collapse unto a unique curve for a given material, up to its peak strength.

The constitutive law for plastic flow is completely specified by the yield surface as defined by equations (7.11) and (7.12), and by the flow rule (7.15), with the definitions (7.16b) and (7.18). Explicit expressions can be given for $(\partial\Phi/\partial\sigma_{ij})$ in (7.15), and for $\dot{\hat{\sigma}}$ in (7.18). They are:

$$\frac{\partial\Phi}{\partial\sigma_{ij}} = \frac{1}{\left(1+\left(\frac{\alpha}{3}\right)^2\right)}\left[\frac{3}{2}\frac{S_{ij}}{\hat{\sigma}} + \frac{\alpha^2}{3}\delta_{ij}\frac{\sigma_m}{\hat{\sigma}}\right] \tag{7.19}$$

$$\dot{\hat{\sigma}} = \frac{1}{\left(1+\left(\frac{\alpha}{3}\right)^2\right)}\left[\frac{3}{2}\frac{S_{ij}}{\hat{\sigma}}\dot{\sigma}_{ij} + \frac{\alpha^2}{3}\frac{\sigma_m}{\hat{\sigma}}\dot{\sigma}_{kk}\right] \tag{7.20}$$

An example of the use of these equations is given in Chapter 9, Section 9.5.

7.3 Postscript

Two final, cautionary, comments. Insufficient experimental data are available to be fully confident about the accuracy of the isotropic hardening law (7.18). There is some evidence that the rate of hardening for hydrostatic compression may be greater than that in simple compression. Indeed, experimental measurements of the hydrostatic and uniaxial compression responses of Alporas and Duocel suggest that the hardening rate is faster for hydrostatic compression (see Figure 7.5 for data presented in the form of true stress versus logarithmic plastic strain curves).

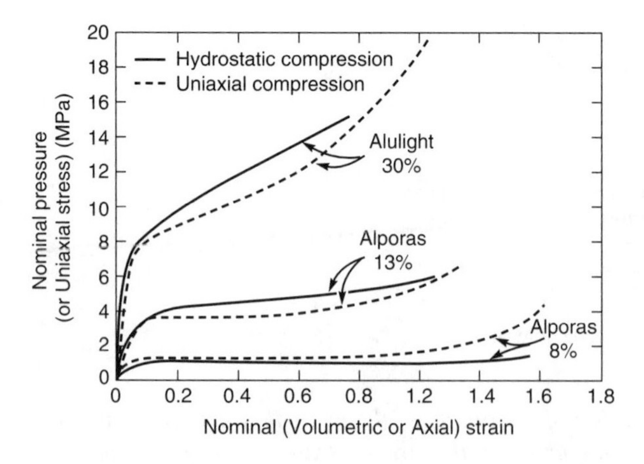

Figure 7.5 *Compression and hydrostatic stress–strain curves for Alporas and Duocel foams: true stress and logarithmic strain*

Some metallic foams harden in compression but soften (and fail prematurely) after yielding in tension. This behavior, not captured by the present constitutive law, is caused by the onset of a new mechanism, that of cell-wall fracture, which progressively weakens the structure as strain increases.

References

Chastel, Y., Hudry, E., Forest, S. and Peytour, C. (1999) In Banhart, J., Ashby, M.F. and Fleck, N.A. (eds), *Metal Foams and Foam Metal Structures*, Proc. Int. Conf. *Metfoam'99*, 14–16 June 1999, MIT Verlag, Bremen, Germany.

Deshpande, V. and Fleck, N.A. (1999) Multi-axial yield of aluminum alloy foams. In Banhart, J., Ashby, M.F. and Fleck, N.A. (eds), *Metal Foams and Foam Metal Structures*, Proc. Int. Conf. *Metfoam'99*, 14–16 June 1999, MIT Verlag, Bremen, Germany.

Deshpande, V.S. and Fleck, N.A. (2000) Isotropic constitutive models for metallic foams. To appear in *J. Mech. Phys. Solids.*

Gibson, L.J. and Ashby, M.F. (1997) *Cellular Solids: Structure and properties*, 2nd edition. Cambridge University Press, Cambridge.

Gioux, G., McCormack, T. and Gibson, L.J. (1999) Failure of aluminum foams under multi-axial loads. To appear in *Int. J. Mech. Sci.*

Miller, R. (1999) A continuum plasticity model of the constitutive and indentation behavior of foamed metals. To appear in *Int. J. Mech. Sci.*

Chapter 8

Design for fatigue with metal foams

In structural applications for metallic foams, such as in sandwich panels, it is necessary to take into account the degradation of strength with cyclic loading. A major cause of this degradation is the nucleation and growth of cracks within the foams. In a closed-cell foam, the cell faces are subject to membrane stresses while the cell edges bend predominantly. Consequently, crack initiation and growth occurs first in the cell faces and then progresses into the cell edges for a closed-cell foam. There is accumulating evidence that an additional fatigue mechanism operates in the cyclic deformation of foams: cyclic creep, also known as ratcheting, under a non-zero mean stress. When a metallic alloy is subjected to cyclic loading with an accompanying non-zero mean stress, the material progressively lengthens under a tensile mean stress, and progressively shortens under a compressive mean stress. Consequently, for a metallic foam, the cell walls progressively bend under a compressive mean stress and progressively straighten under a tensile mean stress. This leads to a high macroscopic ductility in compression, and to brittle fracture in tension.

We shall show later in this chapter that a characteristic feature of metallic foams is their high damage tolerance: the degradation in strength due to the presence of a hole or crack in a foam is usually minor, and there is no need to adopt a fracture mechanics approach. Instead, a design based on net section stress usually suffices. A word of caution, however. It is expected that there should be a critical crack size at which a transition from ductile to brittle behavior occurs for tensile loading and tension–tension fatigue of a notched panel. The precise value of the notch size for which the behavior switches has not yet been determined, but is expected to be large.

8.1 Definition of fatigue terms

First, we need to define some standard fatigue terms. Consider a cylindrical specimen loaded uniaxially by a stress, σ, which varies from a minimum absolute value σ_{min} to a maximum absolute value σ_{max}, as shown in Figure 8.1. For example, for a fatigue cycle ranging from $-1\,\text{MPa}$ to $-10\,\text{MPa}$, we take

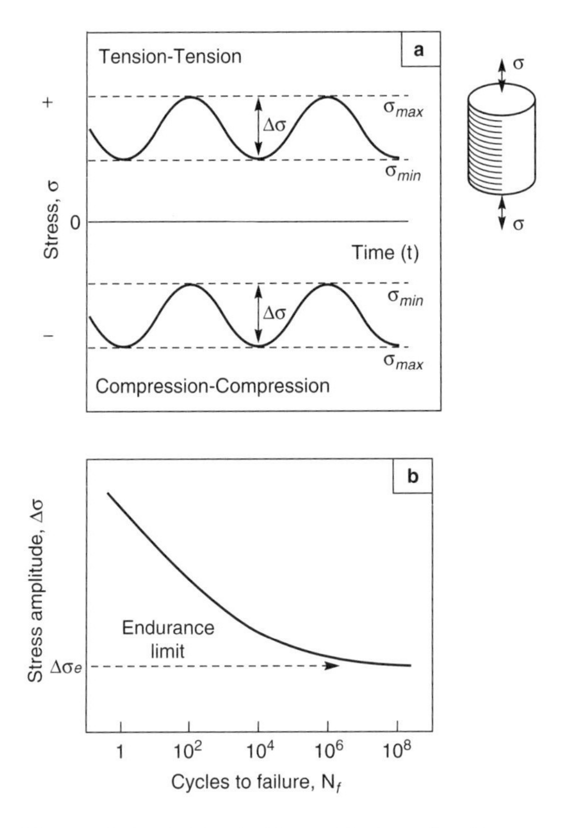

Figure 8.1 *(a) Definition of fatigue loading terms; (b) Typical S–N curve for aluminum alloys, in the form of stress range $\Delta\sigma$ versus number of cycles to failure, N_f. The endurance limit, $\Delta\sigma_e$, is defined, by convention, for a fatigue life of 10^7 cycles*

$\sigma_{min} = 1\,\text{MPa}$ and $\sigma_{max} = 10\,\text{MPa}$. The load ratio R is defined by

$$R \equiv \frac{\sigma_{min}}{\sigma_{max}} \tag{8.1}$$

It is well known that the fatigue life of structural metals such as steels and aluminum alloys is insensitive to the loading frequency, under ambient conditions. This simplification does not hold in the presence of a corrosive medium, such as a hot alkaline solution, or salt water for aluminum alloys. These broad conclusions are expected to hold also for metallic foams.

In low-cycle fatigue testing, the usual strategy is to measure the number of cycles to failure, N_f, for a given constant stress range $\Delta\sigma = \sigma_{max} - \sigma_{min}$, and then to plot the resulting pairs of values $(N_f, \Delta\sigma)$ on log-linear axes. The resulting S–N curve is used in design for finite life (Figure 8.1(b)). Many steels have an S–N curve with a sharp knee below which life is infinite; the corresponding stress range is designated the fatigue limit. This knee is less pronounced for aluminum alloys, and it is usual to assume that 'infinite life' corresponds to a fatigue life of 10^7 cycles, and to refer to the associated stress range $\Delta\sigma$ as the endurance limit, $\Delta\sigma_e$. For conventional structural metals, a superimposed tensile mean stress lowers the fatigue strength, and the knock-down in properties can be estimated conservatively by a Goodman construction: at any fixed life, the reduction in cyclic strength is taken to be proportional to the mean stress of the fatigue cycle normalized by the ultimate tensile strength of the alloy (see any modern text on metal fatigue such as Suresh, 1991, Fuchs and Stephens, 1980 or a general reference such as Ashby and Jones, 1997).

This chapter addresses the following questions:

1. What is the nature of fatigue failure in aluminum alloy foams, under tension–tension loading and compression–compression loading?
2. How does the S–N curve for foams depend upon the mean stress of the fatigue cycle and upon the relative density of the foam?
3. What is the effect of a notch or a circular hole on the monotonic tensile and compressive strength?
4. By how much does a hole degrade the static and fatigue properties of a foam for tension–tension and compression–compression loading?

The chapter concludes with a simple estimate of the size of initial flaw (hole or sharp crack) for which the design procedure should switch from a ductile, net section stress criterion to a brittle, elastic approach. This transition flaw size is predicted to be large (of the order of 1 m) for monotonic loading, implying that for most static design procedures a fracture mechanics approach is not needed and a ductile, net section stress criterion suffices. In fatigue, the transition flaw size is expected to be significantly less than that for monotonic loading, and a brittle design methodology may be necessary for tension–tension cyclic loading of notched geometries.

8.2 Fatigue phenomena in metal foams

When a metallic foam is subjected to tension–tension loading, the foam progressively lengthens to a plastic strain of about 0.5%, due to cyclic ratcheting. A single macroscopic fatigue crack then develops at the weakest section, and progresses across the section with negligible additional plastic deformation. Typical plots of the progressive lengthening are given in Figure 8.2. Shear fatigure also leads to cracking after 2% shear strain.

In compression–compression fatigue the behavior is strikingly different. After an induction period, large plastic strains, of magnitude up to 0.6 (nominal strain measure), gradually develop and the material behaves in a quasi-ductile

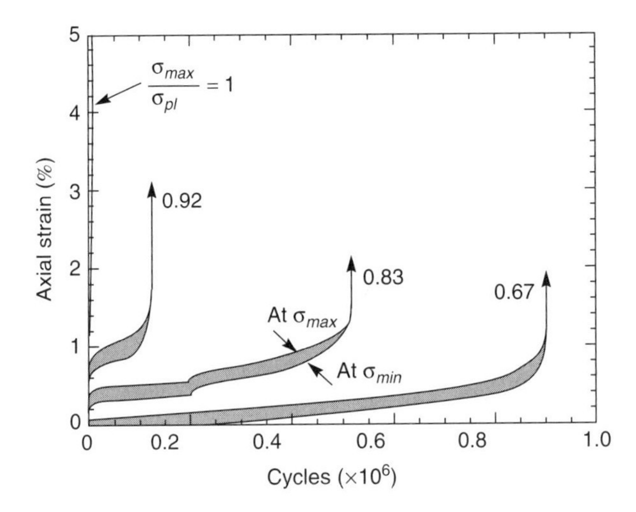

Figure 8.2 *Progressive lengthening in tension–tension fatigue of Alporas, at various fixed levels of stress cycle ($R = 0.1$; relative density 0.11; gauge length 100 mm)*

manner (see Figure 8.3). The underlying mechanism is thought to be a combination of distributed cracking of cell walls and edges, and cyclic ratcheting under non-zero mean stress. Both mechanisms lead to the progressive crushing of cells. Three types of deformation pattern develop:

1. Type I behavior. Uniform strain accumulates throughout the foam, with no evidence of crush band development. This fatigue response is the analogue of uniform compressive straining in monotonic loading. Typical plots of the observed accumulation of compressive strain with cycles, at constant stress range $\Delta\sigma$, are shown in Figure 8.4(a) for the Duocel foam Al-6101-T6. Data are displayed for various values of maximum stress of the fatigue cycle σ_{max} normalized by the plateau value of the yield strength, σ_{pl}.
2. Type II behavior. Crush bands form at random non-adjacent sites, causing strain to accumulate as sketched in Figure 8.3(b). A crush band first forms at site (1), the weakest section of the foam. The average normal strain in the band increases to a saturated value of about 30% nominal strain, and then a new crush band forms elsewhere (sites (2) and (3)), as is sometimes observed in monotonic tests. Type II behavior has been observed for Alporas of relative density 0.08 and for Alcan Al–SiC foams. A density gradient in the loading direction leads to the result that the number of crush bands formed in a test depends upon stress level: high-density regions of the material survive without crushing. Consequently the number of crush bands and the final strain developed in the material increases with increasing stress (Figure 8.4(b)) for an Alcan foam of relative density 0.057.

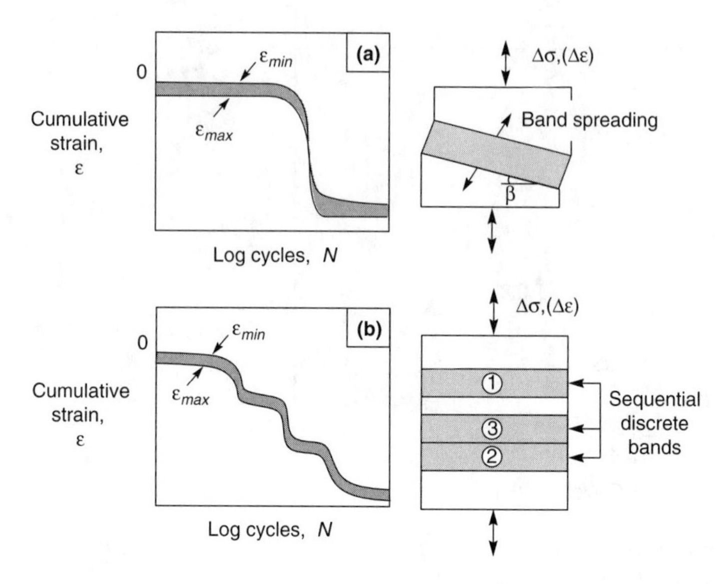

Figure 8.3 *Typical behaviors in compression–compression fatigue of metallic foams at a fixed stress range. (a) Progressive shortening by broadening of a single crush band with increasing cycles; (b) sequential formation of crush bands*

3. Type III behavior. A single crush band forms and broadens with increasing fatigue cycles, as sketched in Figure 8.3(a). This band broadening event is reminiscent of steady-state drawing by neck propagation in a polymer. Eventually, the crush band consumes the specimen and some additional shortening occurs in a spatially uniform manner. Type III behavior has been observed for Alulight of composition Al–1Mg–0.6Si (wt%) and of relative density 0.35, and for Alporas of relative density 0.11. Data for the Alporas are presented in Figure 8.4(c). In both materials, the normal to the crush bands is inclined at an angle of about 20° to the axial direction, as sketched in Figure 8.3(a). The strain state in the band consists of a normal strain of about 30% and a shear strain of about 30%.

A significant drop in the elastic modulus can occur in fatigue, as shown in Figure 8.4(d) for Alporas. This drop in modulus is similar to that observed in static loading, and is a result of geometric changes in the cell geometry with strain, and cracking of cell walls. The precise details remain to be quantified.

A comparison of Figures 8.4(a)–(c) shows that all three types of shortening behavior give a rather similar evolution of compressive strain with the number of load cycles. Large compressive strains are achieved in a progressive manner. We anticipate that this high ductility endows the foams with notch insensitivity in compression–compression fatigue (see Section 8.4 below).

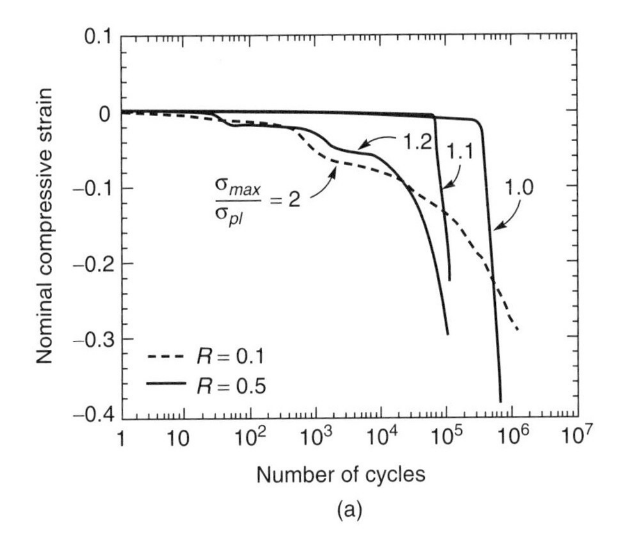

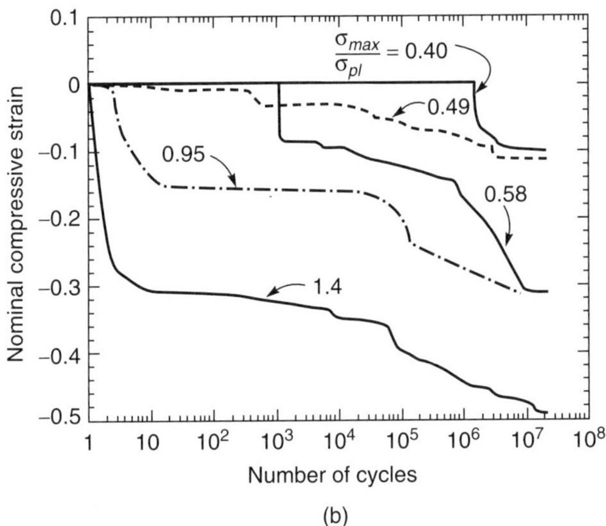

Figure 8.4 *(a) Progressive shortening behavior in compression–compression fatigue for a Duocel Al-6101-T6 foam of relative density 0.08. (b) Progressive shortening behavior in compression–compression fatigue for Alcan foam (relative density 0.057; $R = 0.5$). (c) Progressive shortening behavior in compression–compression fatigue for Alporas foam (relative density 0.11). (d) Progressive shortening behavior in compression–compression fatigue: comparison of the progressive drop in stiffness of Alporas in a monotonic and fatigue test (relative density 0.11)*

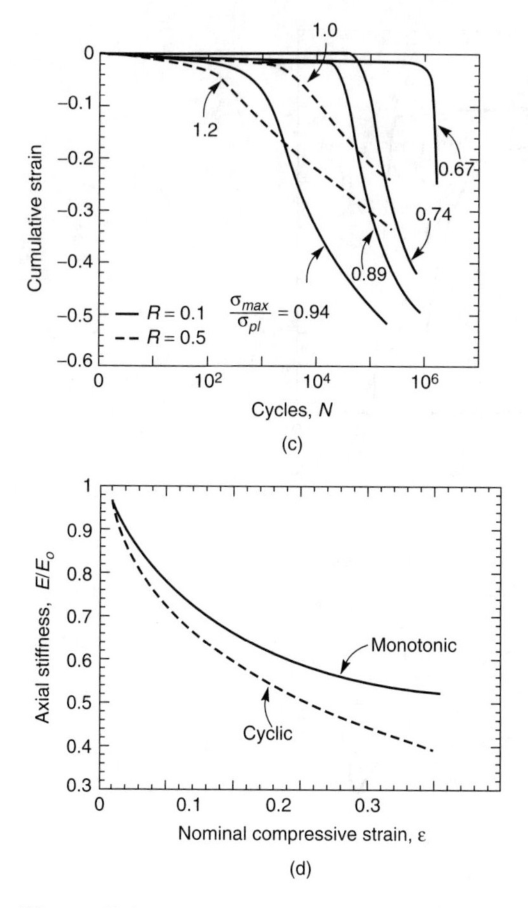

Figure 8.4 *(continued)*

In designing with metal foams, different fatigue failure criteria are appropriate for tension–tension loading and compression–compression loading. Material separation is an appropriate failure criterion for tension–tension loading, while the initiation period for progressive shortening is appropriate for compression–compression loading. Often, a distinct knee on the curve of strain versus cycles exists at a compressive strain of 1–2%, and the associated number of cycles, N_I, is taken as the fatigue life of the material.

8.3 *S–N* data for metal foams

Test results in the form of S–N curves are shown in Figure 8.5 for a number of aluminum alloy foams. Tests have been performed at constant stress range, and

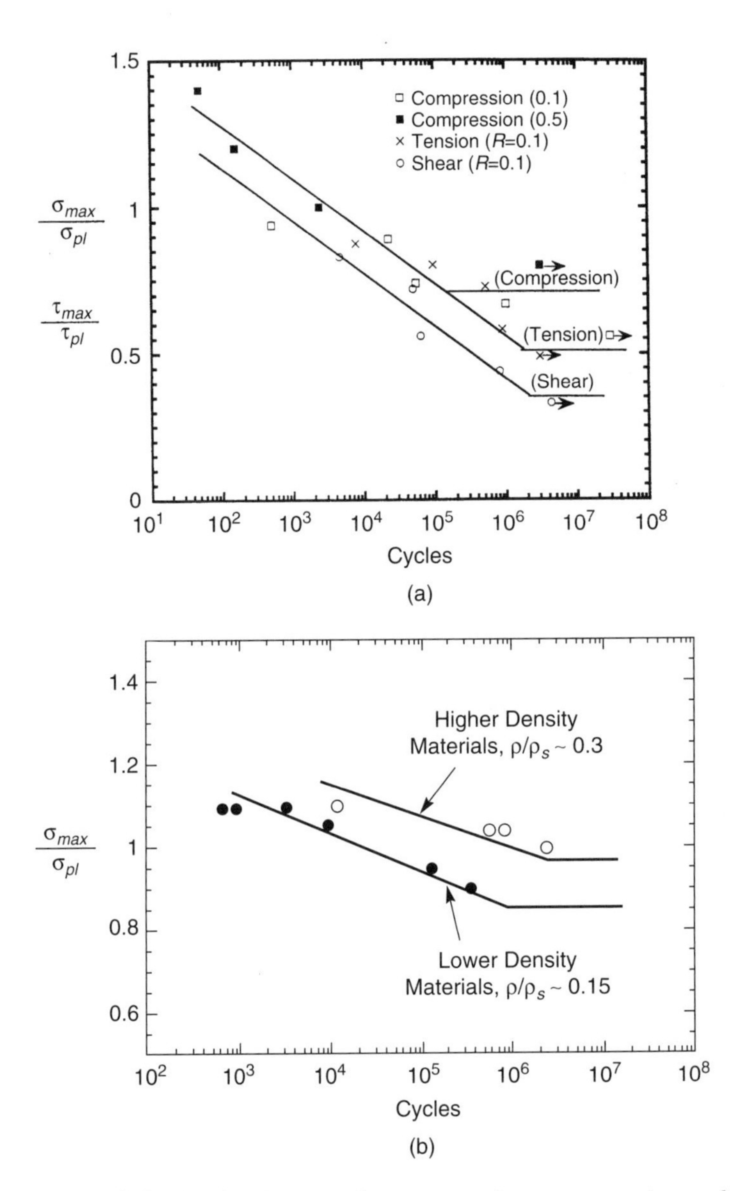

Figure 8.5 *(a) S–N curves for compression–compression and tension–tension fatigue of Alporas foam (relative density 0.11). (b) S–N curves for compression–compression fatigue of Alulight foam ($R = 0.1$). (c) S–N curves for compression–compression and tension–tension fatigue of Alcan foam ($R = 0.1$ and 0.5). (d) S–N curves for compression–compression and tension–tension fatigue of Duocel Al-6101-T6 foam of relative density 0.08*

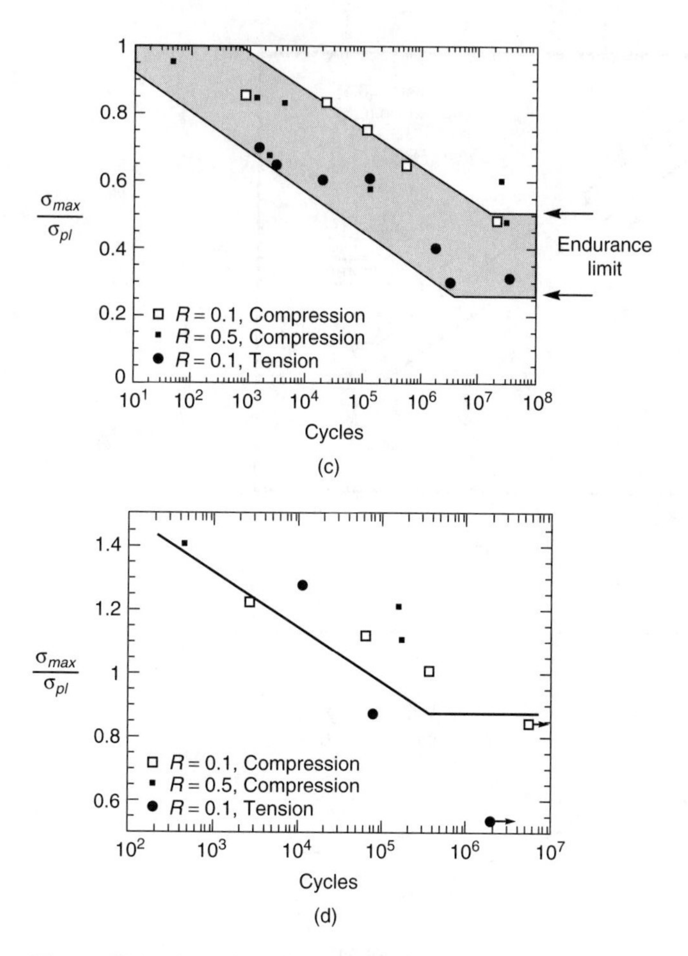

Figure 8.5 *(continued)*

the number of cycles to failure relates to specimen fracture in tension–tension fatigue, and to the number of cycles, N_I, to initiate progressive shortening in compression–compression fatigue. Results are shown in Figure 8.5(a) for an Alporas foam, in Figure 8.5(b) for an Alulight foam, in Figure 8.5(c) for an Alcan foam and in Figure 8.5(d) for a Duocel foam. The following broad conclusions can be drawn from these results:

1. The number of cycles to failure increases with diminishing stress level. An endurance limit can usefully be defined at 1×10^7 cycles, as is the practice for solid metals.
2. The fatigue life correlates with the maximum stress of the fatigue cycle, σ_{max}, rather than the stress range $\Delta\sigma$ for all the foams considered:

compression-compression results for $R = 0.5$ are in good agreement with the corresponding results for $R = 0.1$, when σ_{max} is used as the loading parameter.

3. There is a drop in fatigue strength for tension–tension loading compared with compression–compression fatigue. The fatigue strength is summarized in Figure 8.6 for the various aluminum foams by plotting the value of σ_{max} at a fatigue life of 10^7 cycles versus relative density, over a wide range of mean stresses. The values of σ_{max} have been normalized by the plateau value of the yield strength, σ_{pl}, in uniaxial compression. The fatigue strength of fully dense aluminum alloys has also been added: for tension–tension loading, with $R = 0.1$, the value of σ_{max} at the endurance limit is about 0.6 times the yield strength.

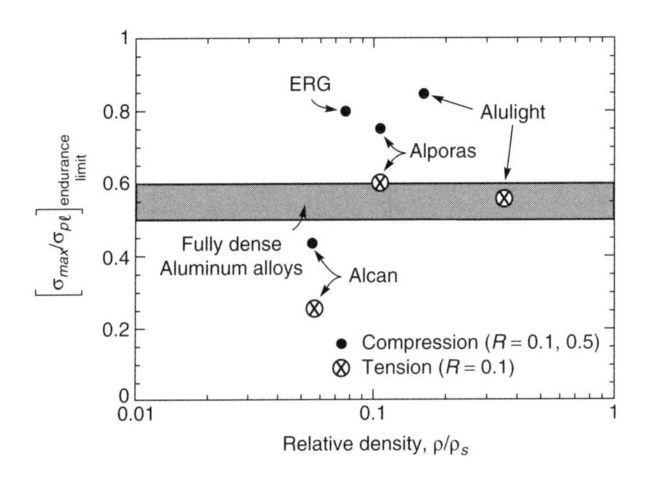

Figure 8.6 *Ratio of σ_{max} at the endurance limit to the monotonic yield strength σ_{pl} for foams, compared with that for tension–tension fatigue of fully dense aluminum alloys at $R = 0.1$*

We conclude from Figure 8.6 that the fatigue strength of aluminum foams is similar to that of fully dense aluminum alloys, when the fatigue strength has been normalized by the uniaxial compressive strength. There is no consistent trend in fatigue strength with relative density of the foam.

8.4 Notch sensitivity in static and fatigue loading

A practical concern in designing with metallic foams is the issue of damage tolerance: in the presence of a notch or hole, does it fail in a notch-insensitive, ductile manner when the net section stress equals the uniaxial strength? Or does it fail in a notch-sensitive, brittle manner, when the local stress at the edge of the hole equals the uniaxial strength? We can answer this question immediately for the case of static compression of a foam containing a circular hole: the large compressive ductility in a uniaxial compression test makes the foam notch insensitive. Experimental results confirm this assessment: a panel of width W, containing a hole of diameter D fails when the net section stress σ_{ns} equals the uniaxial compressive strength σ_{pl} of the foam. On noting that σ_{ns} is related to the applied stress σ^{∞} by $\sigma_{ns} = (1 - (D/W))\sigma^{\infty}$ the net section failure criterion can be rewritten as

$$\sigma^{\infty} = \sigma_{pl}(1 - (D/W)) \tag{8.2}$$

Figure 8.7 confirms that this relation is obeyed for Alporas, Alulight and Alcan aluminum foams, for panels cut in a variety of orientations with respect to the direction of foaming.

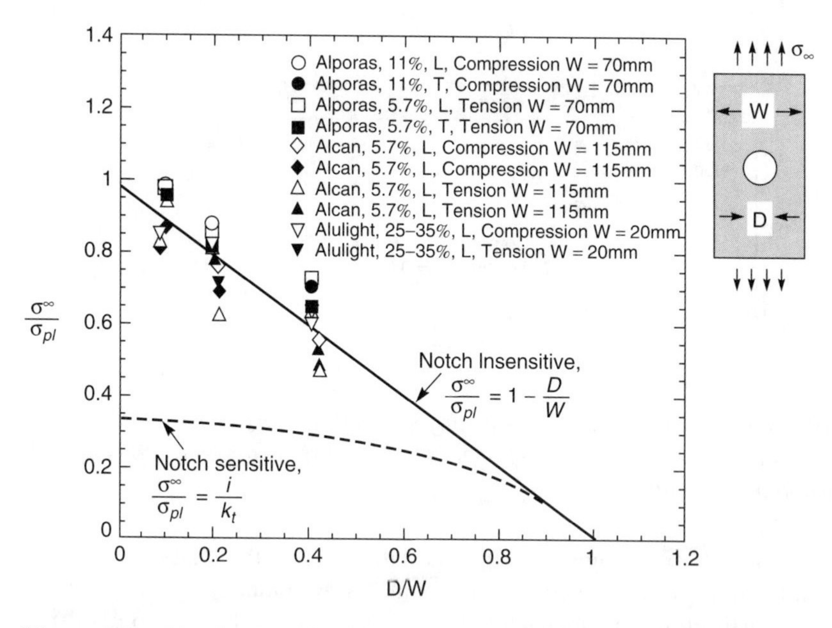

Figure 8.7 *Notch strength of foams, showing behavior to be notch insensitive in both monotonic tension and compression*

For the case of compression–compression fatigue an analogous notched strength criterion applies as follows. Define the endurance limit for a un-notched panel or a notched panel as the gross-section value of σ_{max} in a fatigue test for which progressive shortening does not occur ($N_I > 10^7$ cycles). Then the net section stress criterion reads

$$\sigma_{max,n} = (1 - (D/W))\sigma_{max,un} \tag{8.3}$$

where the subscript n refers to notched, for a hole of diameter D, and the subscript un refers to un-notched specimens. It is reasonable to expect that the net section stress criterion holds in compression–compression fatigue since metal foams progressively shorten under compressive fatigue loading, as Figure 8.4 showed. Notched fatigue data confirm this expectation, as shown in Figure 8.8: aluminum foams are notch-insensitive, and the endurance limit follows the net section stress criterion, given by equation (8.3).

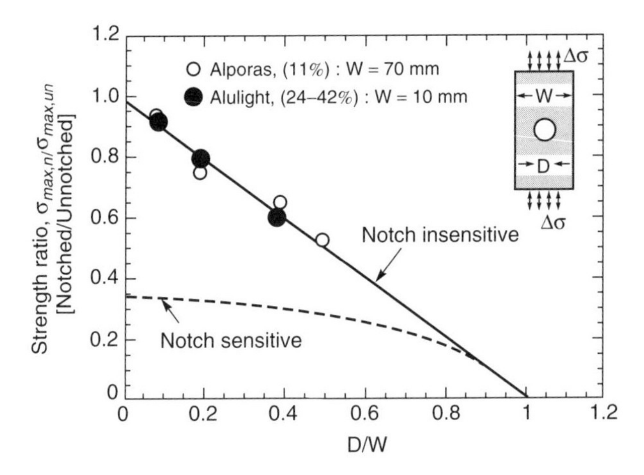

Figure 8.8 *Compression–compression notch fatigue strength of foams, at infinite life; $R = 0.1$*

Now consider a notched metallic foam under monotonic tension. Two types of failure mechanism can be envisaged: ductile behavior, whereby plasticity in the vicinity of the hole is sufficient to diffuse the elastic stress concentration and lead to failure by a net section stress criterion, as given by equation (8.2). Alternatively, a brittle crack can develop from the edge of the hole when the local stress equals the tensile strength of the foam $\sigma_f \approx \sigma_{pl}$. In this case, a notch-sensitive response is expected, and upon assuming a stress concentration K_T for the hole, we expect brittle failure to occur when the remote stress σ^{∞} satisfies

$$\sigma^{\infty} = \sigma_{pl}/K_T \tag{8.4}$$

The following fracture mechanics argument suggests that a transition hole size D_t exists for the foam: for hole diameters D less than D_t the behavior is ductile, with a notched tensile strength σ^{∞} given by equation (8.2). Alternatively, for D greater than D_t the behavior is brittle, with σ^{∞} given by (8.4). A simple micro-mechanical model of failure assumes that the plasticity and tearing of the foam adjacent to the hole can be mimicked by a crack with a constant tensile bridging stress of magnitude σ_{pl} across its flanks, as illustrated in Figure 8.9. This approach follows recent ideas on 'large-scale bridging' of composites, see, for example, the recent review by Bao and Suo (1992). Assume that this bridging stress drops to zero when the crack flanks separate by a critical value δ_0. Measurements of δ_0 using deeply notched specimens reveal that δ_0 is approximately equal to the cell size ℓ for Alporas foam, and we shall make this assumption. This physical picture of the tensile failure of the notched panel is consistent with the notion that the foam has a long-crack toughness of

$$J_c \approx \sigma_{pl}\delta_0 \tag{8.5}$$

and a tensile strength of σ_{pl}. The transition size of hole, D_t, at which the notched strength drops smoothly from σ_{pl} to a value of $\sigma_{pl}/3$ is given by

$$D_t \approx \frac{E\ell}{\sigma_{pl}} \tag{8.6}$$

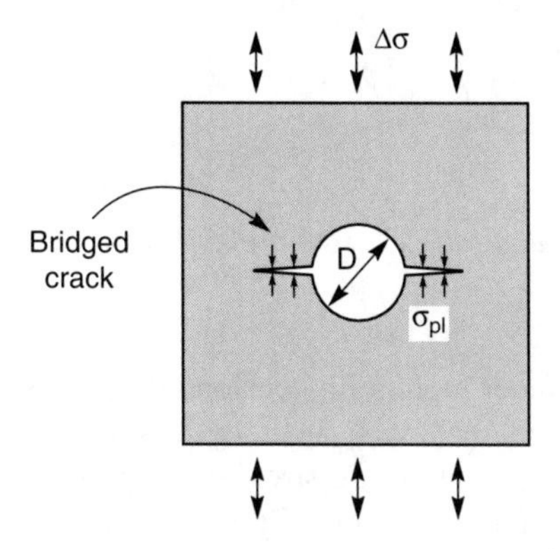

Figure 8.9 *A bridge crack model for yielding and cracking adjacent to an open hole*

The transition hole size is plotted as a function of relative density for a large number of aluminum foams in Figure 8.10. We note that D_t is large, of the order of 1 m. This implies that, for practical problems, the net section stress criterion suffices for notched tensile failure.

The effect of a notch on the tension–tension fatigue strength has not yet been fully resolved. Experiments to date suggest that the net section stress

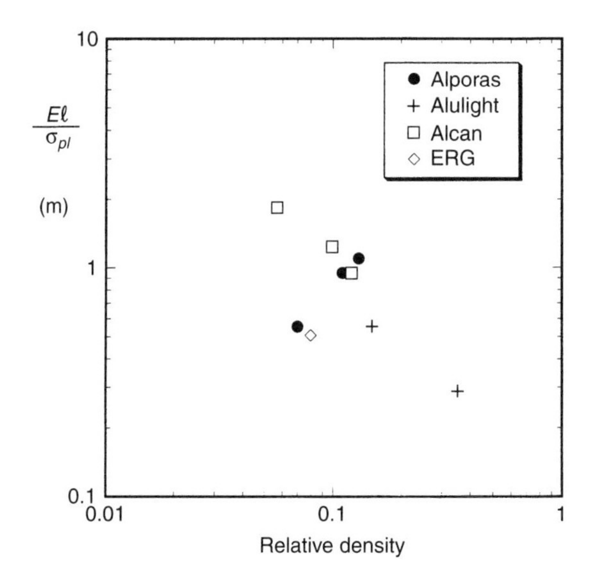

Figure 8.10 *Predicted value of the transition hole size, $D_t \equiv E\ell/\sigma_{pl}$, plotted against relative density*

criterion gives an accurate measure of the drop in fatigue strength due to the presence of a hole. But these tests were done on small holes (maximum diameter equals 30 mm). At the endurance limit, the extent of cyclic plasticity is expected to be much less than the tensile ductility in a static tensile test, and so δ_0 is expected to be much less than the cell size ℓ. Consequently, the magnitude of D_t is expected to be significantly smaller for fatigue loading than for monotonic loading. This effect has been studied for fully dense solids (Fleck *et al.*, 1994) but experiments are required in order to determine the appropriate value for tensile notched fatigue of metallic foams.

References

Ashby, M.F. and Jones, D.R.H. (1997) *Engineering Materials*, Vol. 1, 2nd edition, Butterworth-Heinemann, Oxford.

Banhart, J. (1997) (ed.) *Metallschäume*, MIT Verlag, Bremen, Germany.

Banhart, J., Ashby, M.F. and Fleck, N.A. (eds) (1999) *Metal Foams and Foam Metal Structures*, Proc. Int. Conf. *Metfoam'99*, 14–16 June 1999, Bremen, Germany, MIT Verlag.

Bao, G. and Suo, Z. (1992) Remarks on crack-bridging concepts. *Applied Mechanics Reviews* **45**, 355–366.

Fleck N.A., Kang K.J. and Ashby M.F. (1994) Overview No. 112: The cyclic properties of engineering materials. *Acta Materialia* **42**(2), 365–381.

Fuchs, H.O. and Stephens, R.I. (1980) *Metal Fatigue in Engineering*, Wiley, New York.

Harte, A.-M., Fleck, N.A. and Ashby, M.F. (1999) Fatigue failure of an open cell and a closed cell aluminum alloy foam. *Acta Materialia* **47**(8), 2511–2524.

McCullough, K.Y.G., Fleck, N.A. and Ashby, M.F. (1999). The stress-life behavior of aluminum alloy foams. Submitted to *Fatigue and Fracture of Engng. Mats. and Structures.*

Olurin, O.B., Fleck, N.A. and Ashby, M.F. (1999) Fatigue of an aluminum alloy foam. In Banhart, J., Ashby., M.F. and Fleck, N.A. (eds), *Metal Foams and Foam Metal Structures*, Proc. Int. Conf. *Metfoam'99*, 14–16 June 1999, MIT Verlag, Bremen, Germany.

Schultz, O., des Ligneris, A., Haider, O. and Starke, P (1999) Fatigue behavior, strength and failure of aluminum foam. In Banhart, J., Ashby, M.F. and Fleck, N.A. (eds), *Metal Foams and Foam Metal Structures*, Proc. Int. Conf. *Metfoam'99*, 14–16 June 1999, MIT Verlag, Bremen, Germany.

Shwartz, D.S., Shih, D.S., Evans, A.G. and Wadley, H.N.G. (eds) (1998) *Porous and Cellular Materials for Structural Application*, Materials Reseach Society Proceedings, Vol. 521, MRS, Warrendale, PA, USA.

Sugimura, Y., Rabiei, A., Evans, A.G., Harte, A.M. and Fleck, N.A. (1999) Compression fatigue of cellular Al alloys. *Mat. Sci. and Engineering* **A269**, 38–48.

Suresh, S. (1991) *Fatigue of Materials*, Cambridge University Press, Cambridge.

Zettel, B. and Stanzl-Tschegg, S. (1999) Fatigue of aluminum foams at ultrasonic frequencies. In Banhart, J., Ashby, M.F. and Fleck, N.A. (eds), *Metal Foams and Foam Metal Structures*, Proc. Int. Conf. *Metfoam'99*, 14–16 June 1999, MIT Verlag, Bremen, Germany.

Chapter 9

Design for creep with metal foams

Creep of metals, and of foams made from them, becomes important at temperatures above about one third of the melting point. The secondary, steady-state strain rate $\dot{\varepsilon}$ depends on the applied stress, σ, raised to a power n, typically $3 < n < 6$. The time to failure, t_r, is roughly proportional to the inverse of the steady-state creep rate, that is, $t_r\dot{\varepsilon} \approx C$, where C is a constant. In this chapter the results of models for power-law creep of foams are summarized and the limited experimental data for the creep of metallic foams are reviewed. The creep of metallic foams under multiaxial stress states is described. The results allow the creep of sandwich beams with metallic foam cores to be analysed.

9.1 Introduction: the creep of solid metals

Under constant load, at temperatures T above about one third of the melting temperature T_m, the deformation of metals increases with time t; the material is said to *creep*. The tensile response for a solid metal is shown schematically in Figure 9.1. There are three regimes: primary creep, secondary or steady-state creep and tertiary creep, corresponding, respectively, to decreasing, constant and increasing strain rate. The total creep strain accumulated during the primary regime is usually small compared with that of secondary creep; creep deflections are generally taken as the product of the secondary, steady-state strain rate and the time of loading. In the tertiary creep regime the creep rate accelerates, leading to tensile rupture; the time to failure is taken as the time at which rupture occurs. Engineering design may be limited by either excessive creep deflection or by the time to rupture.

The creep of a metallic foam depends on its relative density and on the creep properties of the solid from which it is made. The dominant mechanism of creep depends on stress and temperature. At low stresses and high temperatures ($T/T_m > 0.8$) diffusional flow along the grain boundaries (at low temperatures) or within the grains (at higher temperatures) can become the dominant mechanism; in this case the steady-state secondary creep rate varies linearly with the applied stress. At higher stresses and more modest temperatures ($0.3 < T/T_m, < 0.8$), climb-controled power-law creep becomes

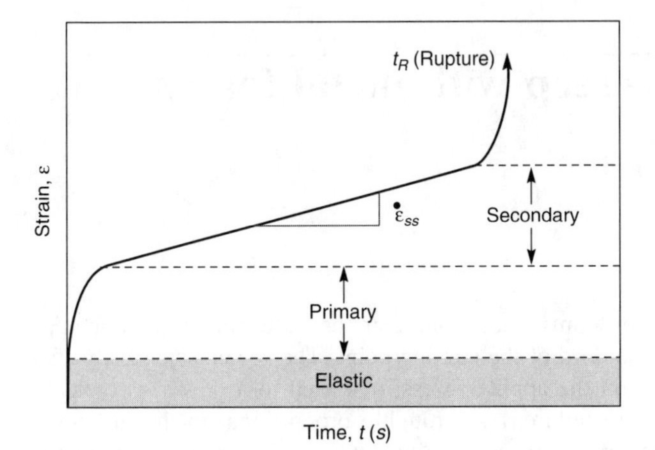

Figure 9.1 *Schematic of creep strain response of a metal under constant load*

dominant; the secondary creep rate, $\dot{\varepsilon}$, then depends on the stress, σ, raised to a power $n > 1$:

$$\dot{\varepsilon} = \dot{\varepsilon}_0 \left(\frac{\sigma}{\sigma_0} \right)^n \tag{9.1}$$

where

$$\dot{\varepsilon}_0 = A \exp\left(-\frac{Q}{RT} \right)$$

Here n, σ_0 and Q are properties of the material (σ_0 is a reference stress and Q is the activation energy for the kinetic process controlling the rate of creep), A has the dimensions 1/second and R is the ideal gas constant (8.314 J/mole K). Typical values for n, σ_0 and Q for power-law creep of several solid metals are listed in Table 9.1.

The time to rupture for a solid metal can be found by assuming that failure is associated with a constant critical strain in the material so that the product

Table 9.1 *Power law creep parameters for solid metals*

Material	n	σ_0 *(MPa)*	Q *(kJ/mole)*
Aluminum	4.4	0.12	142
Nickel	4.6	0.5	284
316 Stainless steel	7.9	33.5	270

of the time to rupture and the secondary strain rate is a constant. The time to rupture, t_r, would then be inversely proportional to the secondary strain rate. In practice, it is found that

$$t_r \dot{\varepsilon}^m \approx C$$

where C is a constant and m is an exponent with a value slightly less than one. Rearranging this expression gives the Monkman–Grant relationship:

$$\log(t_r) + m \log(\dot{\varepsilon}) = \log(C) \tag{9.2}$$

Monkman and Grant found that $0.77 < m < 0.93$ and $-0.48 < \log(C) < -1.3$ for a number of alloys based on aluminum, copper, titanium, iron and nickel (Hertzberg, 1989).

9.2 Creep of metallic foams

The creep strain response of an open-cell Duocel metallic foam is shown in Figure 9.2. In tension, the response is similar to the schematic of Figure 9.1.

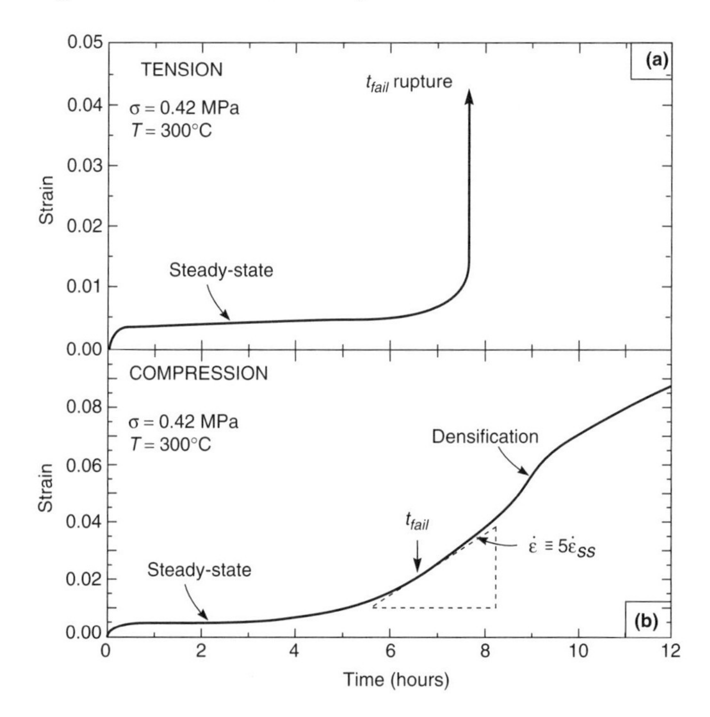

Figure 9.2 *Creep strain plotted against time for open-cell metallic foam under (a) tensile and (b) compressive loading (Duocel open-cell aluminum 6101-T6 foam; Andrews et al., 1999)*

The primary creep regime is short. The extended period of secondary creep is followed by tertiary creep, terminated by rupture. In compression, the behavior is somewhat different. At the end of secondary creep, the strain rate initially increases, but then subsequently decreases; the increase corresponds to localized collapse of a layer of cells. Once complete, the remaining cells, which have not yet reached the tertiary stage, continue to respond by secondary creep at a rate intermediate to the initial secondary and tertiary rates.

For tensile specimens, the time to failure is defined as the time at which catastrophic rupture occurs. For compression loading, the time to failure is defined as the time at which the instantaneous strain rate is five times that for secondary creep.

9.3 Models for the steady-state creep of foams

Open-cell foams respond to stress by bending of the cell edges. If the material of the edges obeys power-law creep, then the creep response of the foam can be related to the creep deflection rate of a beam under a constant load. The analysis is described by Gibson and Ashby (1997) and Andrews *et al.* (1999). The result for the secondary, steady-state creep strain rate, $\dot{\varepsilon}$, of a foam of relative density, ρ^*/ρ_s, under a uniaxial stress, σ, is:

$$\frac{\dot{\varepsilon}}{\dot{\varepsilon}_0} = \frac{0.6}{(n+2)} \left(\frac{1.7(2n+1)}{n} \frac{\sigma^*}{\sigma_0} \right)^n \left(\frac{\rho_s}{\rho^*} \right)^{(3n+1)/2} \tag{9.3}$$

where $\dot{\varepsilon}_0$, n and σ_0 are the values for the solid metal (equation (9.1)). The creep response of the foam has the same activation energy, Q, and depends on stress to the same power, n, as the solid, although the applied stress levels are, of course, much lower. Note that the secondary strain rate is highly sensitive to the relative density of the foam. Note also that this equation can also be used to describe the response of the foam in the diffusional flow regime, by substituting $n = 1$ and using appropriate values of $\dot{\varepsilon}_0$ and σ_0 for the solid.

Closed-cell foams are more complicated: in addition to the bending of the edges of the cells there is also stretching of the cell faces. Setting the volume fraction of solid in the edges to ϕ, the secondary, steady-state creep strain rate of a closed-cell foam is given by:

$$\frac{\dot{\varepsilon}}{\dot{\varepsilon}_0} = \left\{ \frac{\sigma^*/\sigma_0}{\frac{1}{1.7}\left(\frac{n+2}{0.6}\right)^{1/n} \left(\frac{n}{2n+1}\right) \left(\phi \frac{\rho^*}{\rho_s}\right)^{(3n+1)/2n} + \frac{2}{3}(1-\phi)\frac{\rho^*}{\rho_s}} \right\}^n \tag{9.4}$$

When all the solid is in the edges ($\phi = 1$) the equation reduces to equation (9.3). But when the faces are flat and of uniform thickness ($\phi = 0$), it reduces instead to:

$$\frac{\dot{\varepsilon}}{\dot{\varepsilon}_0} = \left(\frac{3}{2}\frac{\rho_s}{\rho^*}\frac{\sigma}{\sigma_0}\right)^n \tag{9.5}$$

9.4 Creep data for metallic foams

The limited creep data for metallic foams are consistent with equation (9.3). Figure 9.3 shows log(steady-state creep rate) plotted against log(stress) and against $1/T$ for a Duocel 6101-T6 aluminum foam, allowing the power, n, and activation energy, Q, to be determined. The measured creep exponent $n = 4.5$

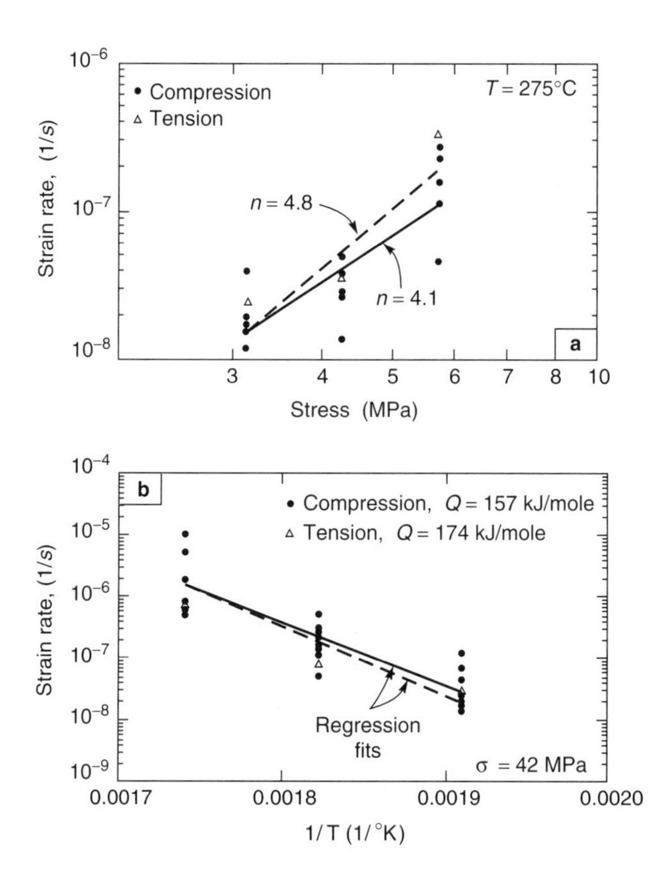

Figure 9.3 *Secondary creep strain rate plotted against (a) stress ($T = 275°C$) and (b) 1/Temperature ($\sigma = 0.42\,MPa$) for an open-cell aluminum foam (Duocel Al 6101 T6 foam, $\rho^*/\rho_s = 0.09$; Andrews et al., 1999)*

is close to the value $n = 4.0$ for solid 6101-T6 aluminum. The measured activation energy, 166 kJ/mole, corresponds well with that for the solid metal ($Q = 173$ kJ/mole). The dependence of the strain rate on foam density is found from a plot of log(steady-state strain rate) against log(relative density), shown in Figure 9.4: the measured power of -6.4 compares well with the expected value of -6.5 from equation (9.3). The steady-state strain rate of the foam is the same in tension and compression. The reference stress $\sigma_0 = 31.6$ MPa.

The results of creep tests on a closed-cell Alporas aluminum foam indicate that it, too, is well described by equation (9.3) at low stresses and temperatures ($\sigma < 0.42$ MPa and $T < 250°C$) (Andrews and Gibson, 1999). At higher stresses and temperatures, the behavior becomes more complicated.

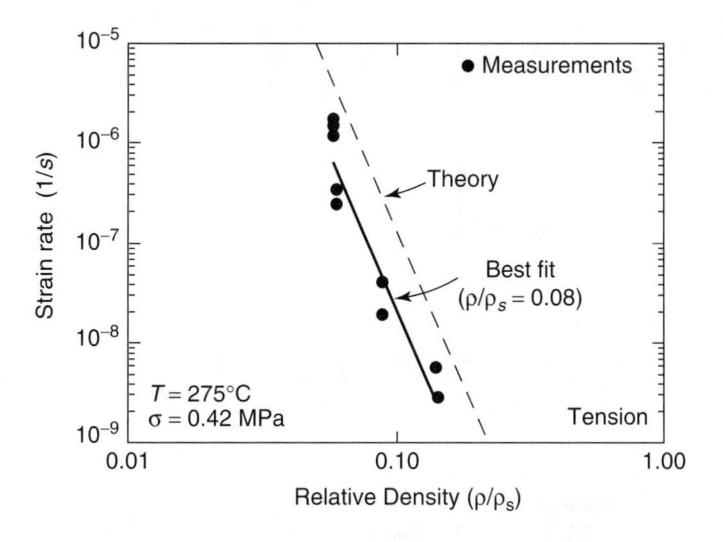

Figure 9.4 *Secondary creep strain rate plotted against relative density at constant stress and temperature for an open-cell foam ($T = 275°C$, $\sigma = 0.42$ MPa, Duocel aluminum 6101-T6 foam; Andrews et al., 1999)*

Figure 9.5 shows the time to failure t_R of a metal foam loaded in tension and compression plotted against the secondary creep strain rate on double-log axes. The value of t_R in compression, based on an instantaneous strain rate of five times the steady-state value, is slightly longer than that in tension. The slopes of the lines give values of the parameter, m, in equation (9.2) of 0.96 and 0.83, for tension and compression, respectively, close to the range found by Monkman and Grant (Hertzberg, 1989). The values of the parameter $\log(C)$ in equation (9.2) are -2.21 and -1.16, meaning that C is smaller than that found by Monkman and Grant, indicating that the aluminum foams have lower creep ductilities than solid aluminum alloys.

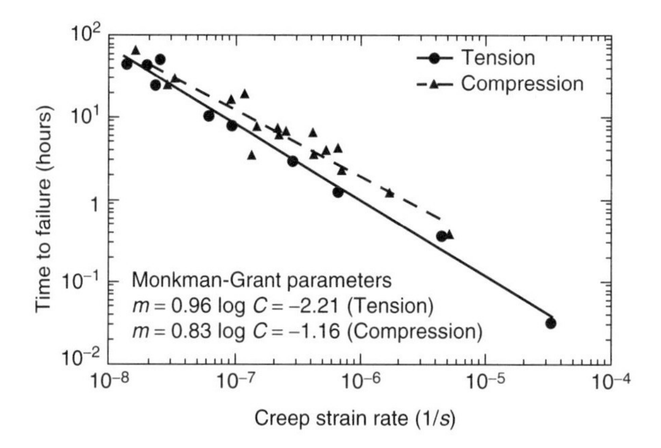

Figure 9.5 *Time to failure in tension and compression plotted against secondary creep strain rate for an open-cell aluminum foam for a range of stresses and temperatures (Duocel aluminum 6101-T6 foam, $\rho^*/\rho_s = 0.09$; Andrews et al., 1999)*

9.5 Creep under multi-axial stresses

The secondary creep rate under multiaxial stresses is found using the constitutive equation for metallic foams (equation (7.12)):

$$\hat{\sigma}^2 = \frac{1}{1 + (\alpha/3)^2}[\sigma_e^2 + \alpha^2 \sigma_m^2] \tag{9.6}$$

and

$$\dot{\varepsilon}_{ij} = f(\hat{\sigma}) \frac{\partial \hat{\sigma}}{\partial \sigma_{ij}} \tag{9.7}$$

In uniaxial tension, we have that $\sigma_e = \sigma_{11}$ and $\sigma_m = \sigma_{11}/3$, giving $\hat{\sigma} = \sigma_{11}$. For uniaxial stress we have that $\partial\hat{\sigma}/\partial\sigma_{11} = 1$ and we know that

$$\dot{\varepsilon}_{11} = \dot{\varepsilon}_0 \left(\frac{\sigma_{11}}{\sigma_0}\right)^n \tag{9.8}$$

giving

$$f(\hat{\sigma}) = \dot{\varepsilon}_0 \left(\frac{\hat{\sigma}}{\sigma_0}\right)^n \tag{9.9}$$

9.6 Creep of sandwich beams with metallic foam cores

A sandwich beam of span ℓ and width b, loaded in three-point bending with a concentrated load, P, is shown in Figure 9.6. The thicknesses of the face

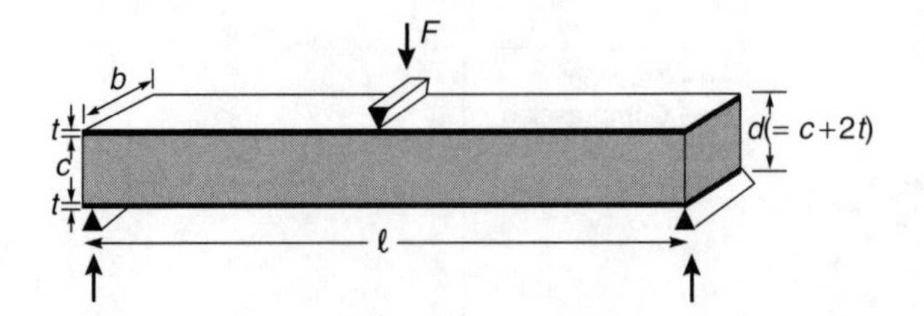

Figure 9.6 *Sandwich beam loaded in three-point bending*

and core are t and c, respectively. The faces have a Young's modulus, E_f, and the core has a Young's modulus, E_c, and a shear modulus G_c. The elastic and plastic deflection of a sandwich beam are analysed in Chapter 10. The creep deflection is found from the sum of the bending and shearing deflection rates. Consider the faces first. The power-law creep response of the faces is given by:

$$\dot{\varepsilon} = A_f \sigma^{n_f} \tag{9.10}$$

where A_f and n_f are the creep parameters of the face material. Assuming that plane sections remain plane,

$$\dot{\varepsilon} = y\dot{\kappa} \tag{9.11}$$

and

$$\sigma = \left(\frac{y\dot{\kappa}}{A_f}\right)^{1/n_f} \tag{9.12}$$

Assuming also that the moment carried by the faces is much larger than that carried by the core, then

$$M = 2\int_{c/2}^{h/2} \left(\frac{y\dot{\kappa}}{A_f}\right)^{1/n_f} yb\,\mathrm{d}y \tag{9.13}$$

$$= B\frac{2b}{A_f^{1/n_f}}\dot{\kappa}^{1/n_f}$$

where

$$B = \frac{1}{2+\dfrac{1}{n_f}}\left[\left(\frac{h}{2}\right)^{2+(1/n_f)} - \left(\frac{c}{2}\right)^{2+(1/n_f)}\right] \tag{9.14}$$

The bending deflection rate at the center of the beam is then:

$$\dot{\delta}_{b\,max} = \left(\frac{P}{4bB}\right)^{n_f} A_f \frac{(\ell/2)^{n_f+2}}{n_f+1}\left(\frac{1}{n_f+2} - 1\right) \tag{9.15}$$

The shear deflection rate is calculated from the creep of the core. The power-law creep of the foam core under uniaxial stress is given by:

$$\dot{\varepsilon} = \dot{\varepsilon}_0 \left(\frac{\sigma}{\sigma_0} \right)^{n_c} \tag{9.16}$$

where $\dot{\varepsilon}_0$, σ_0 and n_c are the creep parameters of the core material. The core is subjected to both normal and shear loading; in general, for metallic foam cores, both are significant. The creep shear strain rate is calculated using equations (9.6) and (9.7):

$$\dot{\varepsilon}_{12} = f(\hat{\sigma}) \frac{\partial \hat{\sigma}}{\partial \sigma_{12}} \tag{9.17}$$

and

$$\hat{\sigma}^2 = \frac{1}{1 + (\alpha/3)^2} [\sigma_e^2 + \alpha^2 \sigma_m^2] \tag{9.18}$$

Noting that

$$\sigma_e = \sqrt{\sigma_{11}^2 + \tfrac{3}{2}\sigma_{12}^2 + \tfrac{3}{2}\sigma_{21}^2}$$

and that $\sigma_m = \sigma_{11}/3$ gives

$$\hat{\sigma} = \left[\sigma_{11}^2 + \frac{\frac{3}{2}}{1 + (\alpha/3)^2} (\sigma_{12}^2 + \sigma_{21}^2) \right]^{1/2} \tag{9.19}$$

Taking the partial derivative with respect to σ_{12} gives:

$$\frac{\partial \hat{\sigma}}{\partial \sigma_{12}} = \frac{\frac{3}{2}\sigma_{12}}{1 + (\alpha/3)^2} \left[\sigma_{11}^2 + \frac{3}{1 + (\alpha/3)^2} \sigma_{12}^2 \right]^{-1/2} \tag{9.20}$$

Using equation (9.9) for $f(\hat{\sigma})$ gives:

$$\dot{\gamma} = 2\dot{\varepsilon}_{12} = \dot{\varepsilon}_0 \left(\frac{\hat{\sigma}}{\sigma_0} \right)^{n_c} \frac{3\sigma_{12}}{1 + (\alpha/3)^2} \left[\sigma_{11}^2 + \frac{3}{1 + (\alpha/3)^2} \sigma_{12}^2 \right]^{-1/2} \tag{9.21}$$

Noting that the normal stress varies through the depth of the beam, in general the shear strain rate has to be integrated over the depth of the beam. The derivative of the shear deflection with respect to x, along the length of the beam, is given by (Allen, 1969):

$$\frac{d\dot{\delta}_s}{\mathrm{d}x} = \dot{\gamma} \frac{c}{d} \tag{9.22}$$

and the shear deflection rate at the center of the beam is:

$$\dot{\delta}_s = \dot{\gamma}\frac{c}{d}\frac{\ell}{2} \tag{9.23}$$

Both normal and shear stresses are, in general, significant for metal foam-core sandwich beams. For a given loading, material properties and geometry, the shear and normal stresses can be evaluated and substituted into equation (9.21) to obtain the shear strain rate. This can then be substituted into equation (9.23) to determine the creep deflection rate of the beam. Preliminary data for creep of sandwich beams are well described by equations (9.23) and (9.21).

References

Allen, H.G. (1969) *Analysis and Design of Structural Sandwich Panels*, Pergamon Press, Oxford.

Andrews, E.W. and Gibson, L.J. (1999) Creep behavior of a closed-cell aluminum foam. *Acta Materialia* **47**, 2927–2935.

Andrews, E.W., Ashby, M.F. and Gibson, L.J. (1999) The creep of cellular solids. *Acta Materialia* **47**, 2853–2863.

Frost, H.J. and Ashby, M.F. (1982) *Deformation Mechanism Maps*, Pergamon Press, Oxford.

Gibson, L.J. and Ashby, M.F. (1997) *Cellular Solids: Structure and Properties*, 2nd edition, Cambridge University Press, Cambridge.

Hertzberg, R.W. (1989) *Deformation and Fracture Mechanics of Engineering Materials*, 3rd edition, Wiley, New York.

Chapter 10

Sandwich structures

Sandwich panels offer high stiffness at low weight. Their cores, commonly, are made of balsa-wood, foamed polymers, glue-bonded aluminum or Nomex (paper) honeycombs. These have the drawbacks that they cannot be used much above room temperature, and that their properties are moisture-dependent. The deficiencies can be overcome by using metal foams as cores. This chapter elaborates the potential of metal-foam-cored sandwich structures.

Competition exists. The conventional way of stiffening a panel is with stringers: attached strips with a profile like a Z, a ⊥ or a top hat. They are generally stiffer for the same weight than sandwich structures, at least for bending about one axis, but they are anisotropic (not equally stiff about all axes of bending), and they are expensive. Metfoam-cored sandwiches are isotropic, can be shaped to doubly curved surfaces, and can be less expensive than attachment-stiffened structures.

Syntactic foams – foams with an integrally shaped skin – offer additional advantages, allowing cheap, light structures to be molded in a single operation. Syntactic polymer foams command a large market. Technologies are emerging for creating syntactic metfoam structures. It is perhaps here that current metal-foam technology holds the greatest promise (see Chapter 16).

10.1 The stiffness of sandwich beams

The first four sections of this chapter focus on the stiffness and strength of sandwich beams and plates (Figure 10.1). In them we cite or develop simple structural formulae, and compare them with experimental results and with the predictions of more refined finite-element calculations. Later sections deal with optimization and compare metfoam-cored sandwiches with rib-stiffened panels.

Consider a sandwich beam of uniform width b, with two identical face-sheets of thickness t perfectly bonded to a metallic foam core of thickness c. The beam is loaded in either in a four-point bend, as sketched in Figure 10.2(a), or a three-point bend as shown in Figure 10.2(b). For both loading cases, the span between the outer supports is ℓ, and the overhang distance beyond the outer supports is H. We envisage that the beams are

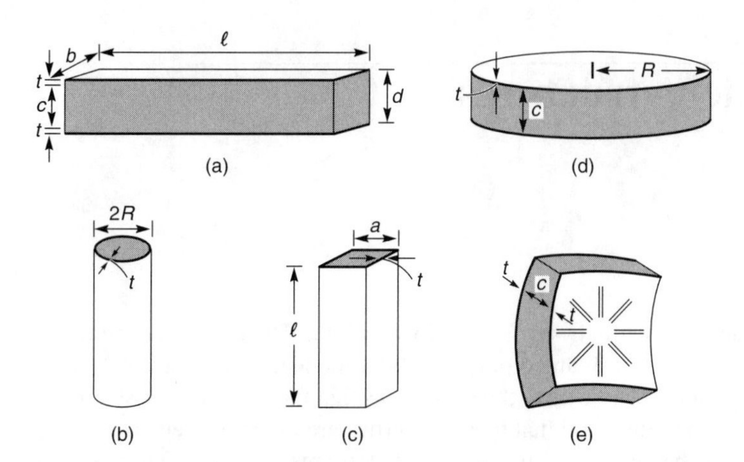

Figure 10.1 *The geometry of sandwich structures: (a) a beam, (b) a circular column, (c) a square column, (d) a circular panel and (e) a shell element*

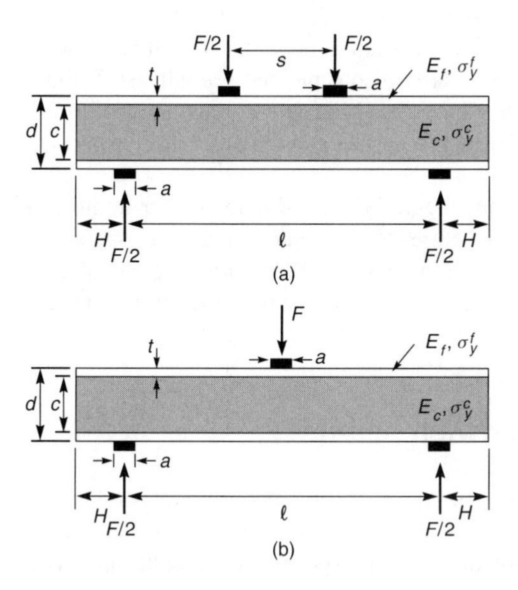

Figure 10.2 *Sandwich beam under (a) four-point bending and (b) three-point bending*

loaded by flat-bottomed indenters of dimension a. The total load applied to the beam is F in each case; for four-point bending the two inner indenters are spaced a distance s apart. Both the core and face-sheets are treated as isotropic, elastic-plastic solids, with a Young's modulus E_f for the face-sheet and E_c for the core.

The elastic deflection δ of the indenters on the top face relative to those on the bottom face is the sum of the flexural and shear deflections (Allen, 1969),

$$\delta = \frac{F\ell^3}{48\,(EI)_{eq}} + \frac{F\ell}{4\,(AG)_{eq}} \tag{10.1}$$

for a three-point bend, and

$$\delta = \frac{F(\ell - s)^2(\ell + 2s)}{48\,(EI)_{eq}} + \frac{F(\ell - s)}{4\,(AG)_{eq}} \tag{10.2}$$

for a four-point bend. Here, the equivalent flexural rigidity $(EI)_{eq}$ is

$$\begin{aligned}(EI)_{eq} &= \frac{E_f btd^2}{2} + \frac{E_f bt^3}{6} + \frac{E_c bc^3}{12} \\ &\approx \frac{E_f btd^2}{2}\end{aligned} \tag{10.3}$$

and $(AG)_{eq}$, the equivalent shear rigidity, is:

$$(AG)_{eq} = \frac{bd^2}{c} G_c \approx bcG_c \tag{10.4}$$

in terms of the shear modulus G_c of the core, the cross-sectional area A of the core, and the spacing $d = c + t$ of the mid-planes of the face-sheets.

The longitudinal bending stresses in the face and core are (Allen, 1969)

$$\sigma^f = \frac{ME_f}{(EI)_{eq}} y \tag{10.5}$$

$$\sigma^c = \frac{ME_c}{(EI)_{eq}} y \tag{10.6}$$

where M is the moment at the cross-section of interest and y is the distance from the neutral axis. The maximum moment is given by

$$M = \frac{F\ell}{4} \tag{10.7}$$

for three-point bending, and by

$$M = \frac{F\,(\ell - s)}{4} \tag{10.8}$$

for four-point bending (here s is the spacing of the inner load-points, Figure 10.2).

10.2 The strength of sandwich beams

In the design of sandwich beams, the strength is important as well as the stiffness. Consider again the sandwich beams under four-point bending and under three-point bending, as sketched in Figure 10.2. Simple analytical formulae can be derived by idealizing the foam core and solid face-sheets by rigid, ideally plastic solids of uniaxial strength σ_y^f and σ_y^c, respectively.

Face yield

When the skins of a sandwich panel or beam are made from a material of low yield strength, face yield determines the limit load F_{fy}. The simplest approach is to assume that plastic collapse occurs when the face sheets attain the yield strength σ_y^f while the core yields simultaneously at a stress level of σ_y^c. For both three- and four-point bending, the collapse load is determined by equating the maximum bending moment within the sandwich panel to the plastic collapse moment of the section, giving

$$F_{fy} = \frac{4bt(c+t)}{\ell}\sigma_y^f + \frac{bc^2}{\ell}\sigma_y^c \tag{10.9}$$

for three-point bending, and

$$F_{fy} = \frac{4bt(c+t)}{\ell - s}\sigma_y^f + \frac{bc^2}{\ell - s}\sigma_y^c \tag{10.10}$$

for four-point bending.

These relations can be simplified by neglecting the contribution of the core to the plastic collapse moment, given by the second term on the right-hand sides of equations (10.9) and (10.10). On labeling this simplified estimate of the collapse load by $\overline{F}_{fy}$ we find that

$$\frac{F_{fy}}{\overline{F}_{fy}} = 1 + \frac{t}{c} + \frac{c}{4t}\frac{\sigma_y^c}{\sigma_y^f} \tag{10.11}$$

for both three- and four-point bending. The error incurred by taking $\overline{F}_{fy}$ instead of F_{fy} is small for typical sandwich panels with a metallic foam core. For example, upon taking the representative values $t/c = 0.1$ and $\sigma_y^c/\sigma_y^f = 0.02$, we obtain an error of 15% according to equation (10.11). It is safer to design on $\overline{F}_{fy}$ than on F_{fy} because a low-ductility core will fracture almost as soon as it yields, causing catastrophic failure.

Indentation

The indentation mode of collapse involves the formation of four plastic hinges within the top face sheet adjacent to each indenter, with compressive yield of the underlying core, as sketched in Figure 10.3. Early studies on the indentation of polymer foams (for example, Wilsea *et al.*, 1975) and more recently on metal foams (Andrews *et al.*, 1999) reveal that the indentation pressure is only slightly greater than the uniaxial compressive strength. The underlying cause of this behavior is the feature that foams compress with little transverse expansion (see Chapter 7 and Gibson and Ashby, 1997).

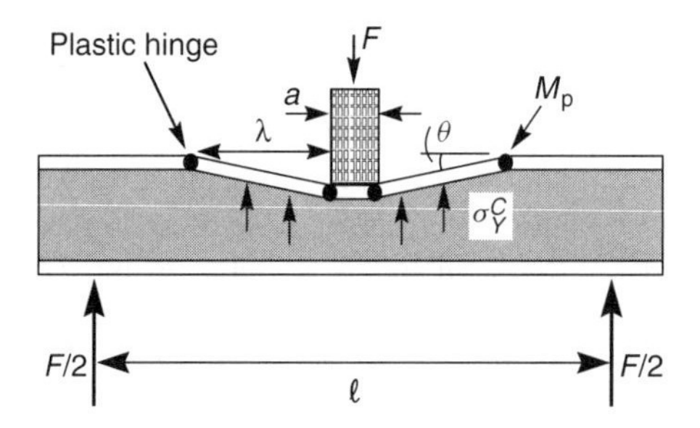

Figure 10.3 *Indentation mode of collapse for a three-point bend configuration*

Consider first the case of three-point bending. Then the collapse load F on the upper indenter can be derived by a simple upper bound calculation. Two segments of the upper face, of wavelength λ, are rotated through a small angle θ. The resulting collapse load is given by

$$F = \frac{4M_p}{\lambda} + (a + \lambda) b\sigma_y^c \tag{10.12}$$

where $M_p = bt^2/4$ is the full plastic moment of the face-sheet section. Minimization of this upper bound solution for F with respect to the free parameter λ gives an indentation load F_I of

$$F_I = 2bt\sqrt{\sigma_y^c \sigma_y^f} + ab\sigma_y^c \tag{10.13}$$

and a wavelength

$$\lambda = t\sqrt{\frac{\sigma_y^f}{\sigma_y^c}} \tag{10.14}$$

We note in passing that the same expression for F_I and λ as given by equations (10.13) and (10.14) are obtained by a lower bound calculation, by considering equilibrium of the face sheet and yield of the face sheet and core. We assume that the bending moment in the face sheets attains a maximum local value of $-M_p$ at the edge of the indenter and also a value of M_p at a distance λ from the edge of the indenter. It is further assumed that the foam core yields with a compressive yield strength σ_y^c and exerts this level of stress on the face sheet, as shown in Figure 10.3. Then, force equilibrium on the segment of a face sheet of length $(2\lambda + a)$ gives

$$F_I = (2\lambda + a)b\sigma_y^c \tag{10.15}$$

and moment equilibrium provides

$$M_p \equiv \tfrac{1}{4}bt^2\sigma_y^f = \tfrac{1}{4}b\lambda^2\sigma_y^c \tag{10.16}$$

Relations (10.15) and (10.16) can be rearranged to the form of (10.13) and (10.14), demonstrating that the lower and upper bound solutions coincide. We conclude that, for a rigid-perfectly plastic material response, these bounds give exact values for the collapse load and for the span length λ between plastic hinges. The presence of two indenters on the top face of the sandwich beam in four-point bending results in a collapse load twice that for three-point bending, but with the same wavelength λ.

Core shear

When a sandwich panel is subjected to a transverse shear force the shear force is carried mainly by the core, and plastic collapse by core shear can result. Two competing collapse mechanisms can be identified, as shown in Figure 10.4 for the case of a beam in three-point bending. Mode A comprises plastic hinge formation at mid-span of the sandwich panel, with shear yielding of the core. Mode B consists of plastic hinge formation both at mid-span and at the outer supports.

Consider first collapse mode A (see Figure 10.4). A simple work balance gives the collapse load F_A, assuming that the face sheets on the right half of the sandwich panel rotate through an angle θ, and that those on the left half rotate through an angle $-\theta$. Consequently, the foam core shears by an angle θ. On equating the external work done $F\ell\theta/2$ to the internal work dissipated within the core of length $(\ell + 2H)$ and at the two plastic hinges in the face sheets, we obtain

$$F_A = \frac{2bt^2}{\ell}\sigma_y^f + 2bc\,\tau_y^c\left(1 + \frac{2H}{\ell}\right) \tag{10.17}$$

where τ_y^c is the shear yield strength for the foam core. Typically, the shear strength of a foam is about two-thirds of the uniaxial strength, $\tau_y^c = 2\sigma_y^c/3$.

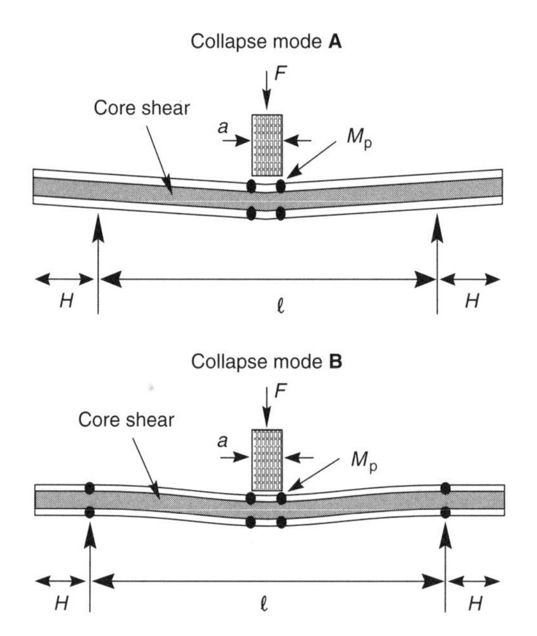

Figure 10.4 *Competing collapse modes A and B for core shear of a sandwich beam in three-point bending*

We note from equation (10.17) that F_A increases linearly with the length of overhang, H, beyond the outer supports.

Second, consider collapse mode B. As sketched in Figure 10.4, this collapse mechanism involves the formation of plastic hinges in the face sheets at both mid-span and at the outer supports. The core undergoes simple shear over the length, L, between the outer supports, with no deformation beyond the outer supports. A work calculation gives for the plastic collapse load F_B,

$$F_B = \frac{4bt^2}{\ell}\sigma_y^f + 2bc\,\tau_y^c \tag{10.18}$$

Since the two calculations given above are upper bounds for the actual collapse load, the lower is taken as the best estimate for the actual collapse load. It is instructive to compare the collapse loads as a function of overhang length H, as sketched in Figure 10.5. Collapse mode A is activated for small lengths of overhang, whereas collapse mode B has the lower collapse load and is activated for large overhangs. The transition length of overhang, H_t, is determined by equating (10.17) and (10.18), giving

$$H_t = \frac{1}{2}\frac{t^2}{c}\frac{\sigma_y^f}{\tau_y^c} \tag{10.19}$$

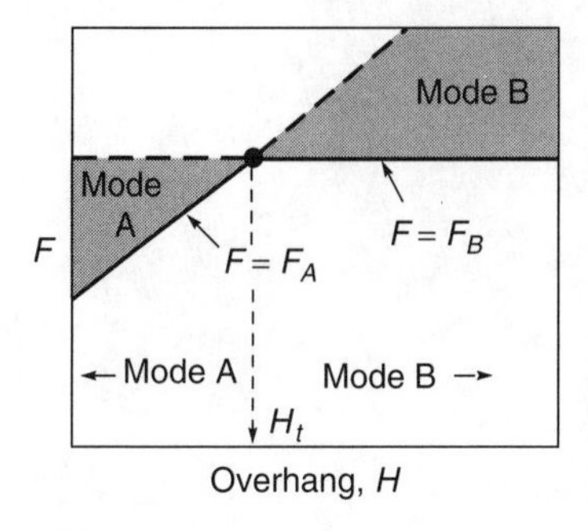

Figure 10.5 *The competition between collapse modes A and B for core shear*

In order to gage the practical significance of the overhang, let us take some representative values for a typical sandwich panel comprising aluminum skins and a metallic foam core, with $c/\ell = 0.1$, $t/c = 0.1$, $\tau_y^c/\sigma_y^f = 0.005$. Then, the transition overhang length, H_t, is given by $H_t = 0.1\ell$: that is, an overhang of length 10% that of the sandwich panel span ℓ is sufficient to switch the collapse mode from mode A to mode B. Furthermore, the enhancement in collapse load due to plastic bending of the face sheets above the load required to shear the core is about 20% for a small overhang, $H \ll H_t$, and is about 40% for $H > H_t$. In much of the current literature on sandwich panels, a gross approximation is made by neglecting the contribution of the face sheets to the collapse load.

Parallel expressions can be derived for the collapse of a sandwich beam in four-point bending by core shear. The collapse load for mode A becomes

$$F_A = \frac{2bt^2}{\ell - s}\sigma_y^f + 2bc\,\tau_y^c\left(1 + \frac{2H}{\ell - s}\right) \tag{10.20}$$

and that for mode B is

$$F_B = \frac{4bt^2}{\ell - s}\sigma_y^f + 2bc\,\tau_y^c \tag{10.21}$$

The transition length of overhang at which the expected collapse mode switches from mode A to mode B is given by the same expression (10.19) as for a beam in three-point bending.

10.3 Collapse mechanism maps for sandwich panels

It is assumed that the operative collapse mechanism for a sandwich beam is the one associated with the lowest collapse load. This can be shown graphically

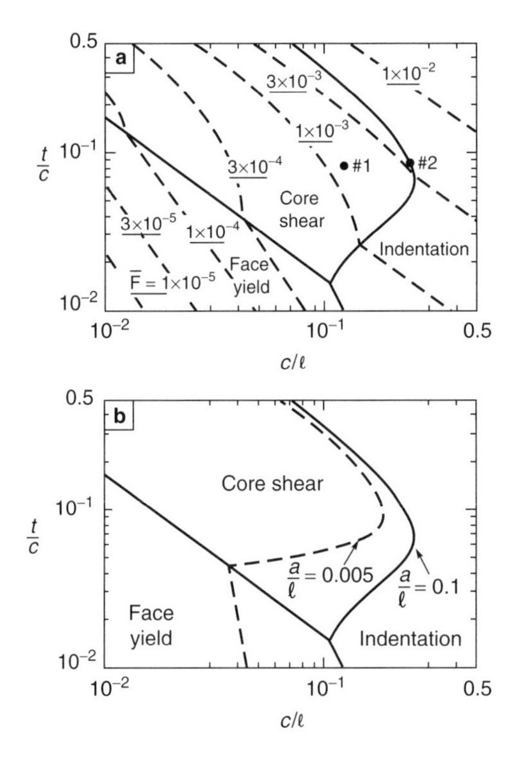

Figure 10.6 *(a) Collapse mechanism map for three-point bending, with flat-bottom indenters. Contours of non-dimensional collapse load* $\overline{F} \equiv F/b\ell\sigma_Y^f$ *are plotted, with* c/ℓ *and* t/c *as axes. The map is drawn for the selected values* $\sigma_Y^c/\sigma_Y^f = 0.005$ *and* $a/\ell = 0.1$*. (b) The effect of the size of the indenter* a/ℓ *upon the relative dominance of the collapse mechanisms for three-point bending*

by plotting a non-dimensional measure of the collapse load $\overline{F} = F/b\ell\sigma_y^f$ on a diagram with the non-dimensional axes c/ℓ and t/c, for selected values of a/ℓ and σ_y^c/σ_y^f. An example is given in Figure 10.6(a) for the case of three-point bending, with $a/\ell = 0.1$ and $\sigma_y^c/\sigma_y^f = 0.005$. It is assumed that the overhang H exceeds the transition value H_t so that core shear is by mode B, as depicted in Figure 10.4. The regimes of dominance for each collapse mechanism are marked: for example, it is clear that failure is by face yield for thin face-sheets (small t/c) and long beams (small c/ℓ). The boundaries of the indentation regime are sensitive to the value taken for a/ℓ: with diminishing a/ℓ the magnitude of the indentation load drops and this regime enlarges on the map, as illustrated in Figure 10.6(b). It is striking that the boundary between the core shear and the indentation regimes has a large curvature, with the indentation mechanism operating at a large values of t/c as well as at small values. This is a consequence of plastic hinge formation within the face sheets in the core shear collapse modes: the collapse load for core shear increases quadratically with increasing t/c due to the contribution from face sheet bending, as seen by examining the first term on the right-hand side of relations (10.17) and (10.18). The contours of collapse load in Figure 10.6(b) show that the load increases along the leading diagonal of the map, with increasing c/ℓ and t/c.

A similar map can be constructed for four-point bending; this is illustrated in Figure 10.7, for the same values $a/\ell = 0.1$ and $\sigma_y^c/\sigma_y^f = 0.005$ as for three-point bending, but with the added parameter $s/\ell = 0.5$. A comparison with the map of Figure 10.6(a) reveals that the domain of face yield shrinks slightly for four-point bending, and indentation almost disappears as a viable mechanism. Core shear dominates the map for the values of parameters selected.

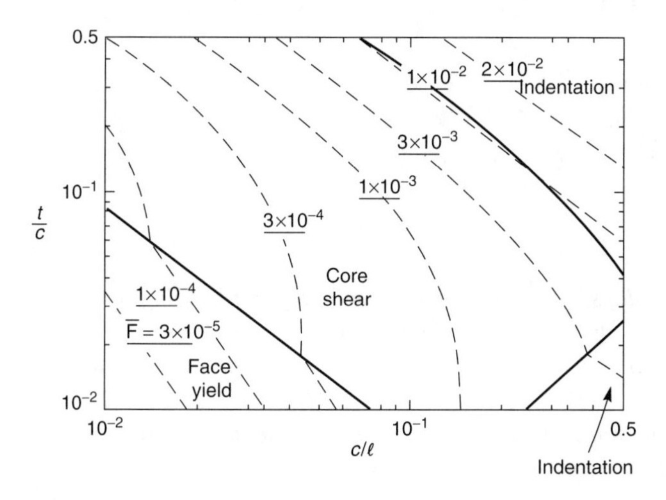

Figure 10.7 *Collapse mechanism map for four-point bending, with flat-bottom indenters. Contours of non-dimensional collapse load* $\overline{F} \equiv F/b\ell\sigma_y^f$ *are plotted for the selected values* $\sigma_y^c/\sigma_y^f = 0.005$, $a/\ell = 0.1$ *and* $s/\ell = 0.5$

Now, some words of caution. The collapse mechanisms described neglect elastic deformation and assume perfectly plastic behavior. Alternative failure modes are expected when the face sheets are made of a monolithic ceramic or composite layers, and behave in an elastic–brittle manner. Then, collapse is dictated by fracture of the face sheets, as analysed by Shuaeib and Soden (1997) and Soden (1996). The above treatment has been limited to the case of flat-bottomed indenters. An alternative practical case is the loading of sandwich beams by rollers, of radius R. This case is more complex because the contact size increases with increasing load. The failure modes of core shear, face yield and indentation have been observed for sandwich beams in a four-point bend with roller-indenters by Harte (1999), and good agreement between the measured strengths of the beams and theoretical predictions are observed, upon making use of the formulae given above but with $a = 0$.

10.4 Case study: the three-point bending of a sandwich panel

A set of three-point bend experiments on foam-cored sandwich panels has been performed by Bart-Smith (1999), using a foam core of Alporas aluminum alloy (relative density $\rho/\rho_s = 0.08$) and 6061-T6 aluminum alloy face sheets. The sandwich beams were loaded by flat-bottom indenters of size $a/\ell = 0.1$; the pertinent geometric and material parameters are summarized in Table 10.1 for two specimen designs, 1 and 2. These designs are based on the collapse load predictions, such that Design 1 sits within the core shear regime, whereas Design 2 lies just within the indentation regime, as marked in Figure 10.6(a).

Table 10.1 *Geometric and material parameters for Designs 1 and 2*

Parameter	*Design 1*	*Design 2*
ℓ, mm	79	40
t, mm	0.8	0.8
c, mm	10	10
b, mm	20	21
H, mm	18	21
a, mm	7.9	4.0
E_f, GPa	70	70
σ_y^f, MPa	263	263
E_c, GPa	0.236	0.263
σ_y^c, MPa	1.5	1.5

The measured load F versus displacement δ curves are shown in Figure 10.8; it was found that the Designs 1 and 2 failed by core shear and by indentation, respectively. For comparison purposes, the predicted elastic stiffness and collapse load have been added to each plot by employing relations (10.1), (10.13) and (10.18). Additionally, the load-deflection responses were calculated by a finite element procedure, using the constitutive model of

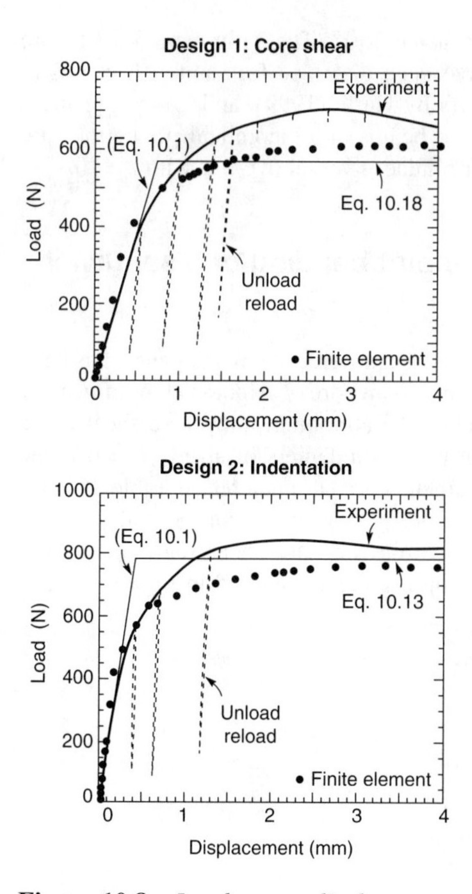

Figure 10.8 *Load versus displacement curves for two designs of sandwich beams in three-point bending compared with the predictions of a finite-element simulation. In the first, the failure mechanism is core shear; in the second, it is indentation. The details of the geometry and material properties are listed in Table 10.1*

Chapter 7. The constitutive model for the foam was calibrated by the uniaxial compressive response, whereas the uniaxial tensile stress–strain response was employed for the solid face sheets. Excellent agreement is noted between the analytical predictions, the finite element calculations and the measured response for both failure modes.

10.5 Weight-efficient structures

To exploit sandwich structures to the full, they must be *optimized*, usually seeking to minimize mass for a given bending stiffness and strength. The next four sections deal with optimization, and with the comparison of optimized sandwich structures with rib-stiffened structures. The benchmarks for comparison are: (1) stringer or waffle-stiffened panels or shells and (2) honeycomb-cored sandwich panels. Decades of development have allowed these to be optimized; they present performance targets that are difficult to surpass. The benefits of a cellular metal system derive from an acceptable structural performance combined with lower costs or greater durability than competing concepts. As an example, honeycomb-cored sandwich panels with polymer composite face sheets are particularly weight efficient and cannot be surpassed by cellular metal cores on a structural performance basis alone. But honeycomb-cored panels have durability problems associated with water intrusion and delamination; they are anisotropic; and they are relatively expensive, particularly when the design calls for curved panels or shells.

In what follows, optimized sandwich construction is compared with conventional construction to reveal where cellular metal sandwich might be more weight-efficient. The results indicate that sandwich construction is most likely to have performance benefits when the loads are relatively low, as they often are. There are no benefits for designs based on limit loads wherein the system compresses plastically, because the load-carrying contribution from the cellular metal core is small. The role of the core is primarily to maintain the positioning of the face sheets.

Structural indices

Weight-efficient designs of panels, shells and tubes subject to bending or compression are determined by structural indices based on load, weight and stiffness. Weight is minimized subject to allowable stresses, stiffnesses and displacements, depending on the application. Expressions for the maximum allowables are derived in terms of these structural indices involving the loads, dimensions, elastic properties and core densities. The details depend on the configuration, the loading and the potential failure modes. Non-dimensional indices will be designated by Π for the load and by ψ for the weight. These will be defined within the context of each design problem. Stiffness indices are defined analogously, as will be illustrated for laterally loaded panels. The notations used for material properties are summarized in Table 10.2. In all the examples, Al alloys are chosen for which $\varepsilon_y^f \equiv \sigma_y^f / E_f = 0.007$.

Organization and rationale

Optimization procedures are difficult to express when performed in a general manner with non-dimensional indices. Accordingly, both for clarity of presentation and to facilitate comprehension, the remainder of this chapter is organized in the following sequence:

Table 10.2 *Material properties*

Property	*Face*	*Foamed core*	*Solid core*
Density (kg/m^3)	ρ_f	ρ_c	ρ_s
Young's modulus (GPa)	E_f	E_c	E_s
Shear modulus (GPa)	–	G_c	–
Yield strength (MPa)	σ_y^f	σ_y^c	σ_y^s
Yield strength in shear (MPa)	–	τ_y^c	–

Notation
b = width
c = core thickness
t = face sheet thickness
ℓ = span length
W = weight
P = load
p = load per unit area

1. A specific example is given for the case of a sandwich plate subject to a uniformly distributed transverse load. This example illustrates issues and procedures related to designs that limit deflections, subject to strength criteria. It then demonstrates how optimization is achieved in terms of a dimensionless load index Π and weight index ψ.
2. Following this example, generalized results are presented for stiffness-limited sandwich beams and plates. These results apply to a range of loadings and give non-dimensional strengths for the local and global weight minima at a specified stiffness.
3. Sandwich panel results are compared with results for waffle-stiffened panels, in order to establish domains of preference.
4. Strength-limited sandwiches are considered and compared with stringer-stiffened construction. Cylindrical sandwich shells are emphasized because these demonstrate clear weight benefits over conventional designs.
5. Overall recommendations regarding sandwich design are given.

10.6 Illustration for uniformly loaded panel

The design of a wide sandwich plate subject to a uniform distributed load will be used to illustrate how the thicknesses of the skins and the cellular metal core are chosen to produce a weight-efficient plate. The structure is shown in Figure. 10.9. The plate is simply supported along its long edges. The span ℓ and the load per unit area p are prescribed. The cellular metal core material is assumed to be pre-selected, such that the analysis can be used to explore the

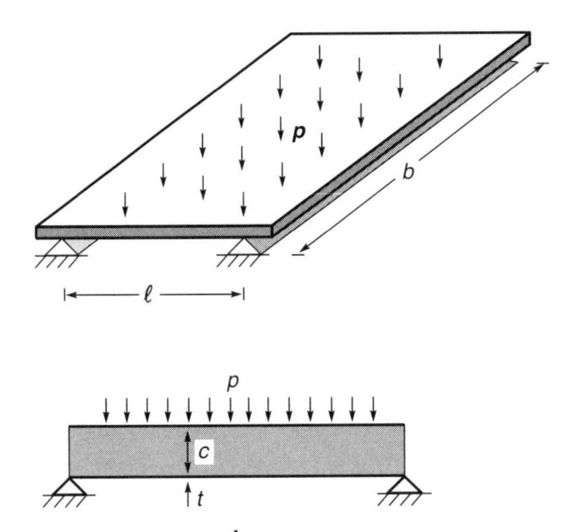

Figure 10.9 *Sandwich panel under uniform load*

effect of different core densities. The objective is to choose the thickness of the core, c, and the thickness of each of the two face sheets, t, so as to minimize the weight of the plate. The width of the plate is assumed large compared with ℓ and the design exercise focuses on a one dimensional, wide plate.

Deflection and failure constraints

The first step in the optimization process is identification of the constraints on failure and deflection. For the sandwich plate under uniform transverse pressure, face sheet yielding and wrinkling must be considered, as well as yielding of the core (Figure 10.10). Deflection constraints must also be imposed to ensure that deflections of an optimally designed panel do not exceed tolerable limits. Normal crushing (indentation) of the core will not be at issue because the loading pressure, p, will necessarily be small compared with the compressive yield strength of the core material.

In the example considered, the face sheet material is the same as that of the fully dense core material (i.e. $\rho_s = \rho_f$, $\sigma_y^s = \sigma_y^f$ and $E_s = E_f$). The relation between the core modulus and density is taken to be

$$E_c/E_s = \alpha_2(\rho_c/\rho_s)^2 \tag{10.22a}$$

where α_2 is a quality factor, with $\alpha_2 = 1$ applying to a core material having relatively inferior properties and $\alpha_2 = 4$ to a material with properties somewhat higher than those currently available. The core is assumed to be isotropic with

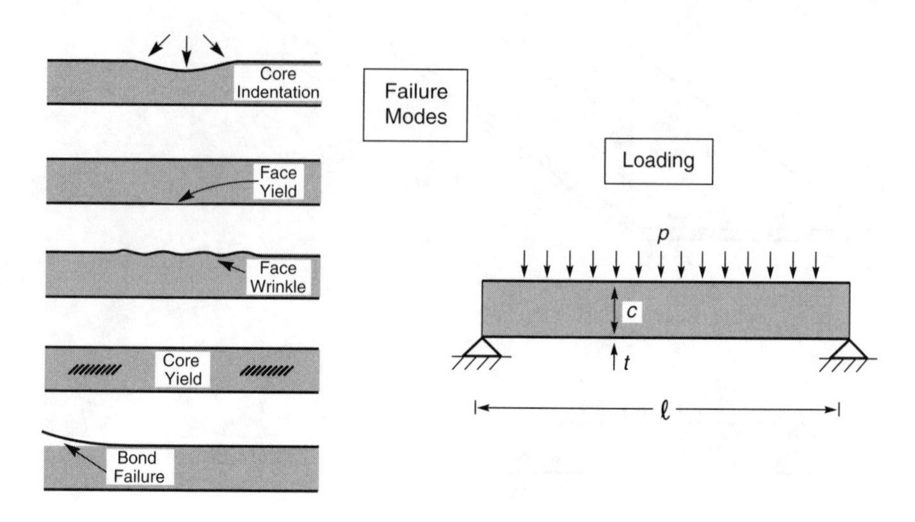

Figure 10.10 *Failure modes in sandwich panels*

$G_c = E_c/2(1 + \nu_c)$. The uniaxial yield strain of the core, σ_y^c/E_c, is assumed to be no less that of the face sheets, σ_y^f/E_f. Consequently, a constraint on face sheet yielding ensures that only shear yielding of the core need be explicitly considered. In this example, the yield strength of the core in shear is taken to be

$$\tau_y^c = \alpha_3(\sigma_y^s/2)(\rho_c/\rho_f)^{3/2} \tag{10.22b}$$

where the quality factor, α_3, is about 0.3.

The weight (in units of kg) of the sandwich plate is

$$W = b\ell(2\rho_f t + \rho_c c) \tag{10.22c}$$

Under the uniform lateral pressure, p, the maximum stress in the face sheets is:

$$\sigma = (p/8)(\ell^2/ct) \tag{10.23a}$$

This maximum occurs at the middle of the plate, with tension on the bottom and compression at the top. The maximum shear stress in the core, which occurs near the supports, is:

$$\tau = (p/2)(\ell/c) \tag{10.23b}$$

The face sheet wrinkling stress is:

$$\sigma_w^f = k(E_f E_c^2)^{1/3} \tag{10.23c}$$

with $k \cong 0.58$ (Allen, 1969). For a given core density, W is to be minimized with respect to t and c subject to three strength-related constraints. These constraints follow directly from the two expressions for the maximum stresses and the wrinkling stress:

$$\text{(face sheet yielding, } \sigma_y^f \geqslant \sigma) \quad ct \geqslant (1/8)\ell^2(p/\sigma_y^f) \tag{10.24a}$$

$$\text{(face sheet wrinkling, } \sigma_w^f \geqslant \sigma) \quad ct \geqslant (1/8)\ell^2\left[p/k(E_f E_c^2)^{1/3}\right] \tag{10.24b}$$

$$\text{(core shear yielding, } \tau_y^c \geqslant \tau) \quad c \geqslant (1/2)\ell(p/\tau_y^c) \tag{10.24c}$$

From the above it is clear that failure is restricted to either face sheet yielding (10.24a) or wrinkling (10.24b), depending on the smaller of σ_y^f and σ_w^f. Note that, with (10.22a), $\sigma_w^f = kE_f{\alpha_2}^{2/3}(\rho_c/p_s)^{4/3}$. Consequently, for an aluminum alloy (Table 10.2) and a core with a quality factor $\alpha_2 = 1$, the wrinkling stress exceeds the yield strength of the face sheets when $\rho_c/\rho_s > 0.036$. We proceed by considering only core materials with relative densities larger than 0.036, with the consequence that face sheet wrinkling (10.24b) is eliminated as a constraint. Were the core density below 0.036, face sheet wrinkling would supersede face sheet yielding.

The deflection δ at the centre of the panel is given by (Allen, 1969; Gibson and Ashby, 1997) (see 10.1).

$$\delta = \frac{2p\ell^4}{B_1 E_f t c^2} + \frac{p\ell^2}{B_2 G_c c} \tag{10.25}$$

where the first term is due to face sheet stretching and the second is the contribution due to core shear. For the present case, the coefficients are $B_1 = 384/5$ and $B_2 = 8$. More results for a selection of loadings and boundary conditions are given in Table 10.3. The maximum deflection will be required to be no greater than $\bar{\delta}$.

Based on the dimensionless load and weight indices,

$$\Pi = p/\sigma_y^f, \qquad \psi = W/(\rho_f b\ell^2) \tag{10.26}$$

the two operative constraints in equation (10.24) can be re-expressed as

$$\text{(face sheet yielding)} \quad (c/\ell)(t/\ell) \geqslant (1/8)\Pi \tag{10.27a}$$

$$\text{(core shear yielding)} \quad (c/\ell) \geqslant (1/2)(\sigma_y^f/\tau_y^c)\Pi \tag{10.27b}$$

The corresponding deflection constraint is

$$\frac{\Pi\left(\sigma_y^f/E_f\right)}{(c/\ell)}\left[\frac{2}{B_1}\left(\frac{\ell}{t}\right)\left(\frac{\ell}{c}\right) + \frac{E_f}{B_2 G_c}\right] \leqslant \frac{\bar{\delta}}{\ell} \tag{10.27c}$$

Table 10.3 *Coefficients for laterally loaded panels*

Loading		B_1	B_2	B_3	B_4
Cantilever	End	3	1	1	1
	Uniform	8	2	2	1
Both ends simply supported	Central	48	4	4	2
	Uniform	384/5	8	9	2
Both ends clamped	Central	192	4	9	2
	Uniform	384	8	12	2

The weight index is

$$\psi = 2(t/\ell) + (\rho_c/\rho_f)(c/\ell) \tag{10.27d}$$

Design diagrams

The remaining constraints on the design variables t and c, expressed by equation (10.27), are plotted as three curves on a design diagram (Figure 10.11(a)), using a particular case wherein the load index is $\Pi = 10^{-4}$ and the maximum allowable deflection is $\bar{\delta}/\ell = 0.02$. The face sheets are assumed to be aluminum and the foam core is also aluminum with $\rho_c/\rho_f = 0.1$ and $\alpha_2 = 1$. Note that the face yielding constraint is independent of the material, whereas that for core shear is independent of the face sheet thickness, but is strongly affected by the core properties.

The relative core thickness cannot lie below the line for core yielding, nor to the left of the curve for face sheet yielding, thereby excluding solutions within the shaded areas of the figure. Similarly, because of the deflection constraint, the solution cannot be to the left of the maximum deflection curve. Consequently, the optimum resides somewhere along the solid curve ADC. (Any combination of c/ℓ and t/ℓ lying to the right of these constraint curves would have a larger weight than the combination with the same value of c/ℓ lying on the closest constraint curve.) Note that, for this example, failure by core yield does not limit the design because the other two are more stringent. That is, the optimum design is limited either by face sheet yielding, above D, or deflection constraint, below D. Evaluating ψ along the two segments AD and DC and then determining its minimum gives the optimum. It is found to reside along DC at location X, where $t/\ell = 0.001$ and $c/\ell = 0.032$. The corresponding weight minimum is obtained from equation (10.6d) as $\psi = 0.00638$. That is, for load index $\Pi = 10^{-4}$, the design is deflection limited. This conclusion changes at larger Π, for reasons explained below.

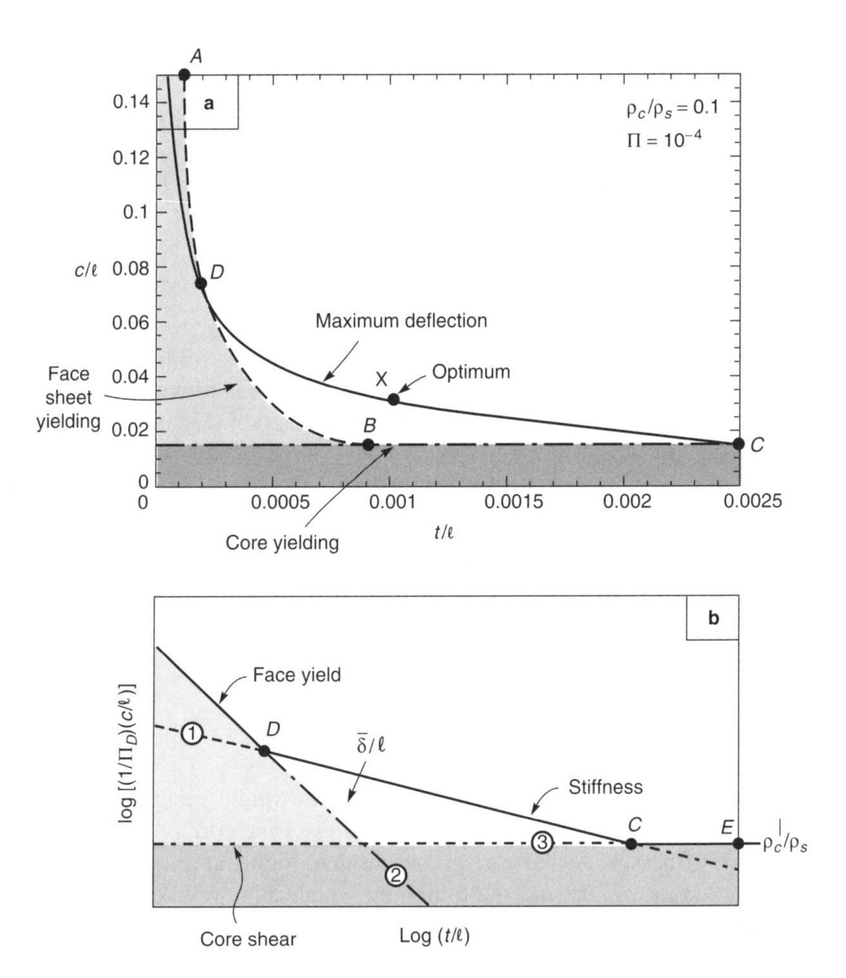

Figure 10.11 *(a) A design map based on panel dimensions for Al alloy sandwich panels at specified load index ($\Pi = 10^{-4}$) and core density ($\rho_c/\rho_s = 0.1$), subject to an allowable displacement ($\bar{\delta}/\ell = 0.02$). (b) A schematic design map using modified coordinates suggested by equation (10.7), showing trends with core properties and allowable stiffness. Line (3) refers to core shear, line (1) to the stiffness constraint and line (2) to face yielding*

The example has been used to illustrate the process used to minimize the weight subject to constraints on failure and deflection. In the present instance, the process can be carried out analytically. Often, however, a straightforward numerical approach is the simplest and most effective way to determine the optimal design. The preceding example may be used to illustrate the numerical methodology:

1. Form a rectangular grid of points $(t/\ell,\ c/\ell)$ covering the potential design space, such as that in Figure 10.11(a).
2. Determine whether each point $(t/\ell,\ c/\ell)$ satisfies all the constraints (10.27): if it does not, reject it; if it does, evaluate ψ.
3. The point which produces the minimum ψ will be close to the optimum design. The grid of points can be further refined if necessary.

Dependence on load index and core density

The above procedures can be used to bring out the regimes within which the design is limited by stiffness or strength. Some assistance is provided by reorganizing equation (10.27) and replotting the design diagram (Figure 10.11(b)). The constraints are:

$$\text{(face yield)}\quad (1/\Pi)(c/\ell) \geqslant (1/8)(t/\ell)^{-1} \tag{10.28a}$$

$$\text{(core shear)}\quad (1/\Pi)(c/\ell) \geqslant (1/2)\left(\sigma_y^f/\tau_y^c\right) \tag{10.28b}$$

$$\text{(deflection)}\quad \bar{\delta}/\ell \geqslant \Pi\left\{\frac{2\sigma_y^f/E_f}{B_1(t/\ell)(c/\ell)^2} + \frac{\sigma_y^f/G_f}{B_2(c/\ell)}\right\} \tag{10.28c}$$

By plotting $(1/\Pi)(c/\ell)$ against t/ℓ using logarithmic axes (Figure 10.11(b)) the respective roles of the relative properties for the core, σ_y^f/τ_y^c and the stiffness constraint, $\bar{\delta}/\ell$, become apparent. The minimum weight design lies along DCE. The precise location depends on the specifics of the core properties and the allowable stiffnesses. As the core properties deteriorate, at given allowable stiffness, line (3) moves up and the minimum weight is now likely to reside along segment CE, being controled by the core. Correspondingly, for a given core material, as the stiffness allowable increases, line (1) displaces downward towards the origin, again causing the minimum weight to reside along CE and be core-controled. Conversely, improvements in the core properties and/or a lower allowable stiffness cause the minimum weight design to reside along DC, such that the panel is stiffness-controled.

Specific results are plotted in Figure 10.12 for four values of the relative core density, with $\bar{\delta}/\ell = 0.01$. Again, the face sheet material and cellular core are taken to be aluminum with $\alpha_2 = 1$. Note that the designs are indeed stiffness-limited at low values of the load index and strength-limited at high load indices. The lower (solid) portion of each curve, below the change in slope, coincides with the former; the two strength constraints being inactive. Moreover, for $\rho_c/\rho_s = 0.3$, the deflection constraint is active over the entire range. At higher load indices (dotted above the change in slope), either or both strength constraints (10.27(a)) and (10.27(b)) are active, with the deflection at the design load being less than $\bar{\delta}$.

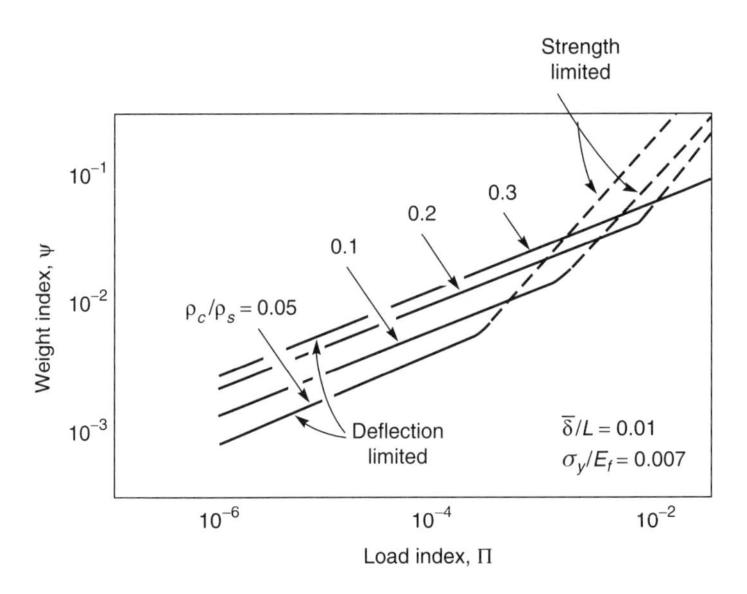

Figure 10.12 *Weight index as a function of load index for optimally designed Al alloy sandwich panels subject to an allowable displacement ($\bar{\delta}/\ell = 0.01$)*

At a low load index, the core with the lowest relative density among the four considered gives the lowest weight structure. At a higher load index, a transition occurs wherein higher core densities produce the lowest weight. Plots such as this can be used to guide the optimal choice of core density.

10.7 Stiffness-limited designs

Sandwich structures

Panels subjected to lateral loads are often stiffness-limited, as exemplified by the panels in the previous section subject to lower load indices. The optimum configuration lies away from the failure constraints and corresponds to a fixed ratio of deflection to load, i.e. to a prescribed stiffness. Stiffness also affects the natural vibration frequencies: high stiffness at low weight increases the resonant frequencies. Minimum weight sandwich panels designed for specified stiffness are investigated on their own merits in this section. The results can be presented in a general form.

The concepts can be found in several literature sources (Allen, 1969; Gerard, 1956; Gibson and Ashby, 1997; Budiansky, 1999). The key results are reiterated to establish the procedures, as well as to capture the most useful results. For laterally loaded flat panel problems, non-dimensional coefficients (designated B_i) relate the deflections to the applied loads. Details are given in Section 6.3. The key results are repeated for convenience in Table 10.3. Their use will be illustrated throughout the following derivations.

An analysis based on laterally loaded sandwich beams demonstrates the procedure. The results are summarized in Figure 10.13, wherein the weight is minimized subject to prescribed stiffness. The construction of this figure, along with the indices, will be described below. The global weight minimum has the core density ρ_c as one of the variables. This minimum is the lower envelope of the minimum weight curves obtained at fixed ρ_c: three of which are shown.

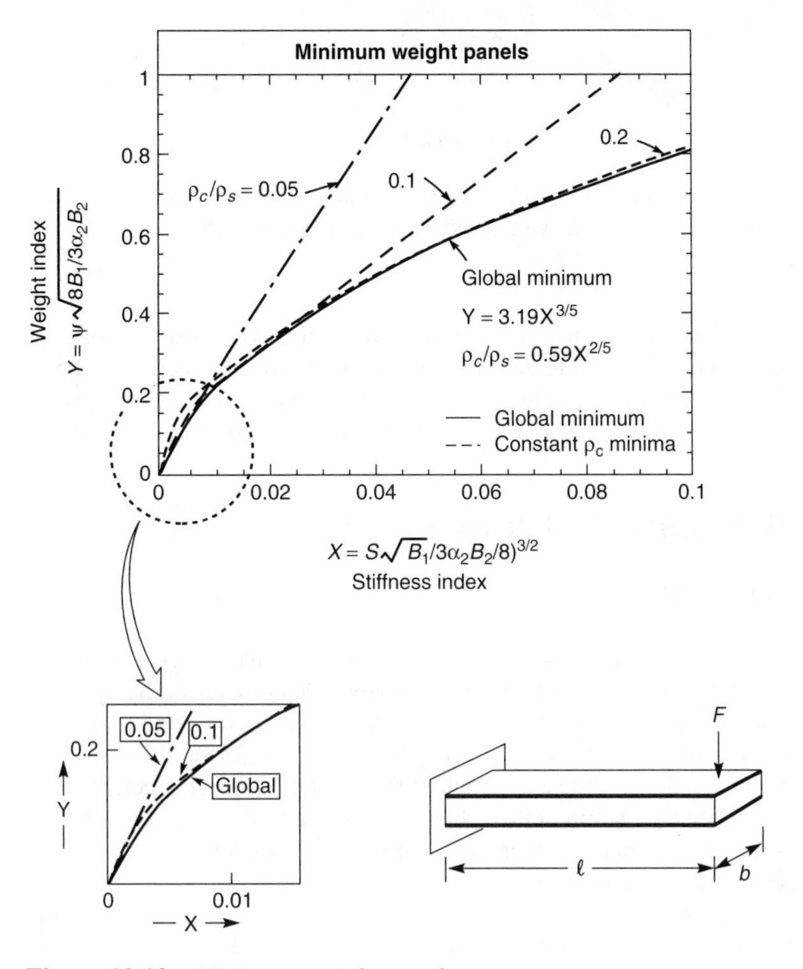

Figure 10.13 *Minimum weight panels*

To construct such plots, the stiffness S and weight W are first defined. With δ as the deflection and P as the transverse load amplitude, the compliance per unit width of panel can be obtained directly from equation (10.25) as (Allen, 1969):

$$\frac{1}{S} \equiv \frac{b}{P/\delta} = \frac{2\ell^3}{B_1 E_f t c^2} + \frac{\ell}{B_2 c G_c} \tag{10.29}$$

The first term is the contribution from face sheet stretching and the second from core shear. The result applies to essentially any transverse loading case with the appropriate choice for B_1 and B_2 (Table 10.3). The weight index is:

$$\psi = \frac{W}{\rho_f \ell^2 b} = \frac{2t}{\ell} + \frac{\rho_c}{\rho_f}\frac{c}{\ell} \tag{10.30}$$

These basic results are used in all subsequent derivations.

The global minimum

In the search for the *global optimum*, the free variables are t, c, and ρ_c, with due recognition that G_c in equation (10.29) depends on ρ_c. If the core density ρ_c is taken to be prescribed so that G_c is fixed, the optimization proceeds by minimizing ψ with respect to t and c for specified stiffness. Inspection of equation (10.29) reveals that the most straightforward way to carry out this process is to express t in terms of c, allowing equation (10.29) to be re-expressed with c as the only variable. For this problem, the expressions are sufficiently simple that the minimization can be carried out analytically: the results are given below. In other cases, the minimization may not lead to closed-form expressions. Then, the most effective way to proceed is to create a computer program to evaluate ψ (or W itself) in terms of c and to plot this dependence for specified values of all the other parameters over the range encompassing the minimum. Gibson and Ashby (1997) have emphasized the value of this graphical approach which can be extended to consider variations in core density simply by plotting a series of curves for different ρ_c, analogous to what was done in Figure 10.12.

If the global minimum is sought, the dependence of the shear modulus of the core must be specified in terms of its density. In the following examples, the material comprising the face sheets is assumed to be the same as the parent material for the core ($E_f = E_c = E_s$, when $\rho_c = \rho_s = \rho_f$). The dependence of Young's modulus of the core on ρ_c is again expressed by equation (10.22a). Taking the Poisson's ratio of the cellular core material to be $\frac{1}{3}$ (Gibson and Ashby, 1997) then $G_c/E_f = (\frac{3}{8})(\rho_c/\rho_s)^2$.

Although it seems paradoxical, the search for the global optimum gives rise to simpler expressions than when the core density is fixed. The result of minimizing ψ with respect to t, c and ρ_c with S prescribed is:

$$\frac{c}{\ell} = 2\left[\frac{18\alpha_2 B_2 S}{B_1^2 E_f}\right]^{1/5} \qquad \frac{t}{\ell} = \frac{B_1}{96\alpha_2 B_2}\left(\frac{c}{\ell}\right)^3 \qquad \frac{\rho_c}{\rho_s} = 8\frac{t}{c} \tag{10.31}$$

Note that t, c and ρ_c are explicitly defined at the global minimum. It is readily verified that the globally optimized beam has the following two characteristics: (1) the compliance, $S^{-1} = 6\ell^3/(B_1 E_f tc^2)$, has exactly twice the contribution from the core as from the face sheets (the second term in equation (10.29) is twice the first term); (2) perhaps more surprisingly, the weight of the core is exactly four times that of the combined weight of the two face sheets.

At the minimum, equation (10.31) enables the weight index, Y, to be expressed in terms of the stiffness index X,

$$Y = \tfrac{5}{16}(48X)^{3/5} \tag{10.32a}$$

where

$$Y = \left(\frac{8B_1}{3\alpha_2 B_2}\right)^{1/2} \psi \quad \text{and} \quad X = \frac{B_1^{1/2}}{(3\alpha_2 B_2/8)^{3/2}}\frac{S}{E_f} \tag{10.32b}$$

These are the two non-dimensional quantities plotted as the global minimum in Figure 10.13. They contain all the information needed to characterize the support and load conditions encompassed by Table 10.3, inclusive of the coefficient determining the stiffness of the cellular metal, α_2. The core density at the global minimum can also be expressed as a function of X:

$$\rho_c/\rho_s = (48X)^{2/5}/8 \tag{10.33}$$

Fixed core density

Return now to the minimization of weight with ρ_c fixed. This is done for the same class of sandwich beams: parent core material the same as the face sheet material. Minimization of weight at prescribed stiffness now relates c and t to a free parameter ξ such that,

$$\frac{c}{\ell} = \left[\frac{3\alpha_2 B_2(\rho_c/\rho_s)^3 \xi}{4B_1}\right]^{1/2} \quad \text{and} \quad \frac{t}{c} = \frac{(1-2\xi)(\rho_c/\rho_s)}{4} \tag{10.34}$$

Each value of ξ generates a minimum weight beam for the fixed core density, with the stiffness specified by the index X, defined in equation (10.11), now given by:

$$X = \frac{16\sqrt{2}(\rho_c/\rho_s)^{5/2}\xi^{3/2}}{1-4\xi^2} \tag{10.35}$$

The associated non-dimensional weight index defined in equation (10.11) is

$$Y = 2\sqrt{2}(\rho_c/\rho_s)^{3/2}\xi^{1/2}\frac{(3-2\xi)}{(1-2\xi)} \tag{10.36}$$

Curves of Y as a function of X are included in Figure 10.13 for three values of the relative core density, ρ_c/ρ_s. Each curve necessarily lies above the global optimum, touching that curve only where its core density happens to coincide with that of the globally optimized sandwich beam. Note, however, that the minimum weight beams with fixed core density exhibit substantial stiffness and weight ranges over which they are close to optimal. For example, the beams with $\rho_c/\rho_s = 0.2$ have weights which are only slightly above the global minimum over the stiffness range, $0.02 < X < 0.1$.

The results presented in Figure 10.13 have the merit that they are universal, encompassing a variety of support and load conditions through the dimensionless axes. The graphical approach (Gibson and Ashby, 1997) could have been used to produce the same results for specific sets of support and load conditions. For problems with greater geometric complexity, the graphical approach might well be the most effective way to seek out the lowest weight designs.

Failure limits

Application of these weight diagrams is limited by the occurrence of the various failure modes sketched in Figure. 10.10: yielding either of the face sheets or in the core, and face wrinkling. These phenomena govern the maximum load at which the beam responds elastically. The illustration in Section 10.6 has elaborated the consequences. Some additional considerations are given in this section.

Reiterating from equation (10.24a), *face yielding* commences when the maximum tensile or compressive stress caused by bending reaches the yield strength, σ_y^f. For a given set of support and loading conditions, the maximum stress in the face sheet is determined by the coefficient B_3 in Table 10.3, such that yielding in the face sheets commences when the transverse load satisfies

$$P \geqslant \frac{bB_3ct}{\ell}\sigma_y^f \tag{10.37}$$

This condition can be re-expressed in terms of the stiffness index X defined in equation (10.32) and superimposed on Figure 10.13. That is, the condition that the face sheets of the globally optimal beam remain elastic will be violated if

$$X \leqslant \frac{1}{48}\left\{\left(\frac{16B_1}{\alpha_2 B_2 B_3}\right)\left(\frac{P}{b\ell\sigma_y^f}\right)\right\}^{5/4} \tag{10.38}$$

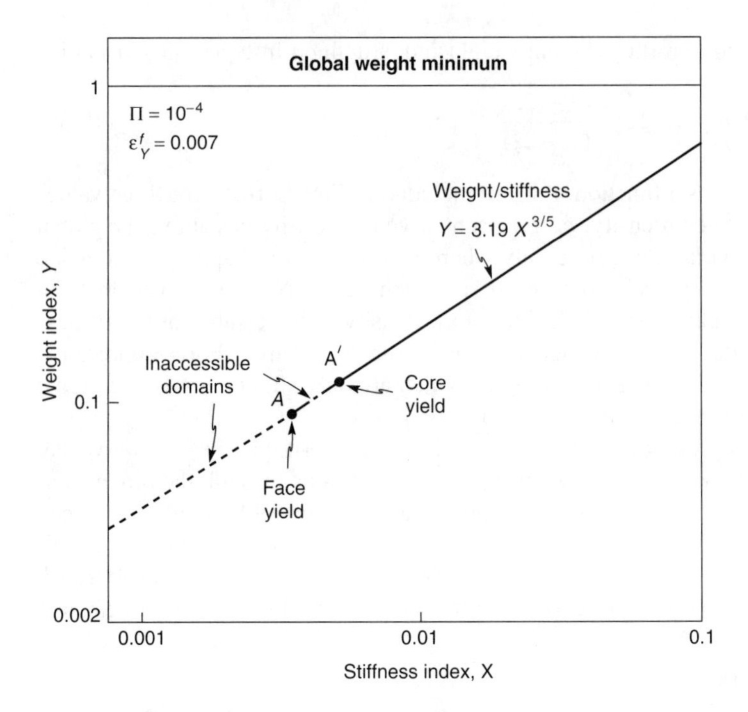

Figure 10.14 *The relationship between X and Y showing the region bounded by core and face yield*

For specific conditions, the equality of (10.38) corresponds to a point on the curve of weight versus stiffness (Figure 10.14) below which the elastic predictions are no longer valid. For values of the stiffness index below this point, weights in excess of the global minimum would be needed to ensure that the beam remains elastic. Logarithmic axes have been used in Figure 10.14 to highlight the inadmissible range.

Analogous conditions exist for *core yielding* (equation 10.24(c)). The coefficient B_4 is defined such that the maximum shear stress in the core is $P/(B_4bc)$, and thus yielding occurs when

$$P > B_4 bc\tau_y^c \tag{10.39}$$

This condition can also be written in terms of X for the case of the globally optimized beams. Core yielding invalidates the results based on the elastic optimization if

$$X \leqslant \frac{1}{48}\left\{\frac{1}{B_4}\left(\frac{B_1}{\alpha_2 B_2}\right)^{1/2}\left(\frac{P}{b\ell\tau_y^c}\right)\right\}^5 \tag{10.40}$$

Elastic wrinkling of the face sheets may also occur (equation (10.24b)).

Note that, as the stiffness index increases, the face sheet thicknesses needed to achieve minimum weights increase substantially, relative to core thickness and density. Consequently at lower stiffnesses yielding is more likely to intervene because of the thinner face sheets and lower core densities at the global weight minimum (equation (10.31)). For yielding to be avoided, the loads on the structure must by limited by equations (10.37) and (10.39).

Stiffened panels

The principal competitors for sandwich systems subject to biaxial bending are waffle-stiffened panels (Figure 10.15). For comparison, it is convenient to re-express the result for the globally optimized sandwich beam (equation (10.32)) in the form:

$$\frac{W}{b\ell^2} = \rho_f \left(\frac{P/\delta}{bE_f}\right)^{3/5} \left[\frac{15}{B_1^{1/5}(18\alpha_2 B_2)^{2/5}}\right] \tag{10.41}$$

For a waffle panel subject to bending about one of the stiffener directions, the weight and stiffness are related by:

$$\frac{W}{b\ell^2} = \frac{72}{5}\left(\frac{P/\delta}{E_0 B_1 b}\right)\left(\frac{\ell}{d_s}\right)^2 \tag{10.42}$$

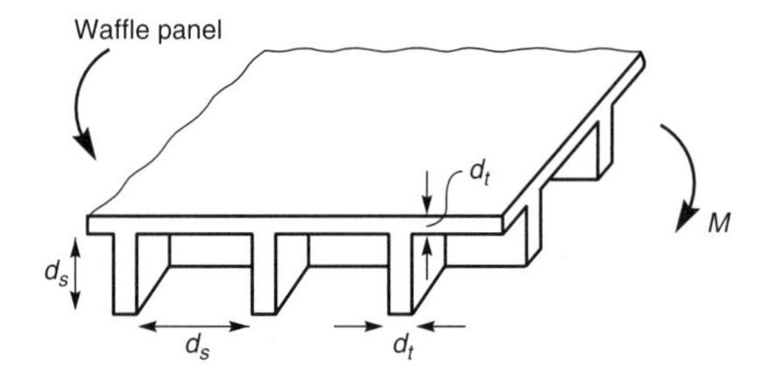

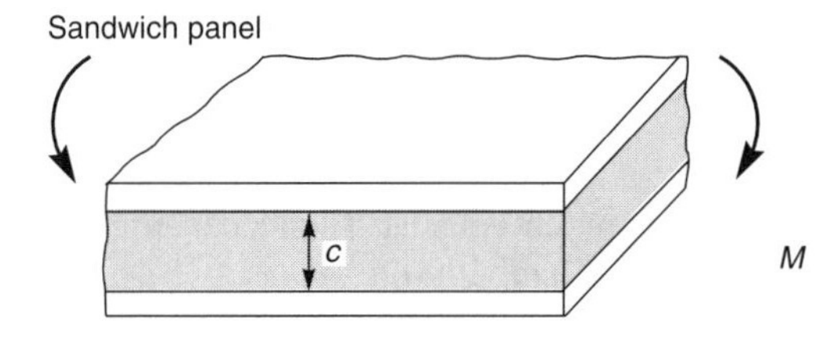

Figure 10.15 *A waffle-stiffened panel and a sandwich panel loaded in bending*

where the web depth, d_s, is defined in Figure 10.15, with E_0 as Young's modulus for the material comprising the panel. At *equal weights of the sandwich and waffle panels*, the web depth of the waffle is

$$\frac{d_s}{\ell} = \sqrt{\frac{72}{125}\left(\frac{\rho_0}{\rho_f}\right)\left(\frac{E_f}{E_0}\right)\left(\frac{18\alpha_2 B_2}{B_1^2}\right)^{1/5}}\left(\frac{P/\delta}{bE_f}\right)^{1/5} \quad (10.43)$$

Comparison with the result for c/ℓ in equation (10.31) for the globally optimized sandwich panel at equivalent weights gives:

$$\frac{d_s}{c} = \frac{\sqrt{6}}{5}\sqrt{\left(\frac{\rho_0}{\rho_f}\right)\left(\frac{E_f}{E_0}\right)} \quad (10.44)$$

This result is stiffness independent because the sandwich and waffle panels have the identical functional dependence. Accordingly, a waffle panel made from the same material as a sandwich panel ($\rho_f = \rho_0$, $E_f = E_0$) has a slightly smaller overall thickness at the same weight and stiffness. The choice between sandwich and waffle panels, therefore, depends primarily on manufacturing cost and durability.

10.8 Strength-limited designs

Cylindrical shells

Strength-limited sandwich structures can be weight competitive with stiffener-reinforced designs (the lowest weight designs in current usage). Shells are a more likely candidate for sandwich construction than axially compressed panels or columns because both hoop and axial stresses are involved, enabling the isotropy of sandwich panels to be exploited. There are two basic requirements for sandwich shells: (1) sufficient core shear stiffness for adequate buckling strength, (2) sufficiently large yield strength of the metal foam to maintain the buckling resistance of the shell, particularly in the presence of imperfections. Numerical methods are needed to determine minimum weights of both sandwich and stringer-reinforced configurations. Some prototypical results for a cylindrical shell under axial compression illustrate configurations in which sandwich construction is preferred.

General considerations

The perfect cylindrical shell buckles axisymmetrically at a load per circumferential length, N, given by (Tennyson and Chan, 1990)

$$\frac{N}{E_f R} = \frac{2tc}{\sqrt{1-\nu_f^2}R^2}\left[1 - \frac{\mu}{\sqrt{2}}\right] \quad (10.45a)$$

where

$$\mu = \frac{E_f t}{\sqrt{2(1 - \nu_f^2) G_c R}} \tag{10.45b}$$

Here, R is the shell radius, t the face sheet thickness, c the core thickness. The parameter μ measures the relative shear compliance of the core. It must be less than $1/\sqrt{2}$ to avoid localized shear kinking of the shell wall. This result (10.45) applies if the length of the shell ℓ is at least several times the axial buckle wavelength. It also assumes that sufficiently strong end support conditions are in effect. End conditions modify (10.45) slightly, but not as much as imperfections, to be discussed later.

The condition for face yielding of the perfect shell prior to (or simultaneous with) buckling is:

$$N = 2t\sigma_y^f \tag{10.46}$$

Yielding of the core does not directly affect the load-carrying capacity since it supports no significant load. However, it will affect the ability of the core to suppress face sheet wrinkling and to maintain the shear stiffness necessary for post-buckling load-carrying capacity. Yielding of the core prior to buckling is avoided if the axial strain in the unbuckled shell ($\varepsilon = N/(2tE_f)$) does not exceed the uniaxial compressive yield strain of the core. In other words, core yielding is excluded if its yield strain is larger than that of the face sheets.

If neither the core nor the face sheets yield, face sheet wrinkling may occur. It is governed by the condition introduced earlier (equation (10.24b)). In terms of the load per unit circumferential length, the *onset of face sheet wrinkling* in the perfect shell occurs when

$$N = 2tk(E_f E_c^2)^{1/3} \tag{10.47}$$

For shells with core parent material identical to the face sheet material ($\rho_s = \rho_f$ and $E_f = E_s$), with core modulus and density still related by equation (10.22a), the weight index is

$$\psi = \frac{W}{2\pi R^2 \ell \rho_f} = 2\frac{t}{R} + \frac{\rho_c}{\rho_s}\frac{c}{R} \tag{10.48}$$

When the shell buckles elastically, uninfluenced by yielding or wrinkling, global optimization of the shell could be carried out analytically to obtain the values of t, c and ρ_c which minimize the weight of the shell for prescribed N. This elastic optimization implies that the core weight is twice that of the two face sheets. However, it is of little practical value because face sheet yielding or wrinkling invariably intervene at the levels of load index wherein the sandwich shell is weight competitive.

Fixed core density

Consider shells with prescribed core density ρ_c designed to carry load per circumferential length, N. Subject to the inequality, $\sigma_y^f/E_f < k\alpha_2^{2/3}(\rho_c/\rho_s)^{4/3}$, core yielding always excludes face sheet wrinkling, and vice versa. An optimal design having face sheet yielding coincident with buckling is obtained by using equation (10.46) to give t, and then using that expression in equation (10.45) to obtain c. The weight follows from equation (10.27). The procedure for design against simultaneous wrinkling and buckling follows the same steps, but now using equation (10.47) rather than (10.46).

The outcome from the above optimization is shown in Figure 10.16(a) in the form of a plot of the weight index as a function of load index, $N/(E_f R)$, for two values of core density. These plots are constructed for a core with stiffness at the low end of the range found for commercial materials ($\alpha_2 = 1$). The core has been assumed to remain elastic, and the yield strength of the face sheets is that for a structural aluminum alloy. The range of the load index displayed is that for which sandwich cylinders have a competitive edge over more conventional construction comprising axial stiffeners. Included in Figure 10.16(a) is the weight-optimized shell with hat-shaped axial stiffeners which buckles between rings spaced a distance R apart.

The sandwich results are independent of shell length, whereas the axially stiffened results do depend on the segment length, typically R. The optimized sandwich shells in Figure 10.16(a) experience simultaneous buckling and face sheet yielding, except in the range $N/(E_f R) \leqslant 4 \times 10^{-6}$ where the buckling is elastic. Note that the relative weight of the core to the total weight for these shells (Figure 10.16(b)) is very different from that predicted by the elastic global analysis. For shells in the mid-range of the structural index, the core weight comprises only about 25% of the total.

This example affirms that metal foam core sandwich shells can have a competitive advantage over established structural methods of stiffening, particularly at relatively low structural indices.

Global minimum

To pursue the subject further, the sandwich shells have been optimized with respect to relative core density ρ_c, as well as t and c, allowing for all possible combinations of face sheet yielding and wrinkling. Simultaneously, the consequence of using a core with superior stiffness is addressed by assuming a core having properties comparable to the best commercial materials ($\alpha_2 = 4$ rather than $\alpha_2 = 1$). The results for the fully optimized foam-core sandwich shells are plotted in Figure 10.17 with accompanying plots for the optimal relative density of the core. The operative deformation modes in the optimal design are indicated in the plot of core density. Wrinkling and buckling are simultaneous at the lowest values of the load index; wrinkling, yielding and buckling in the intermediate range; and yielding and buckling at the high end. Again, the global elastic

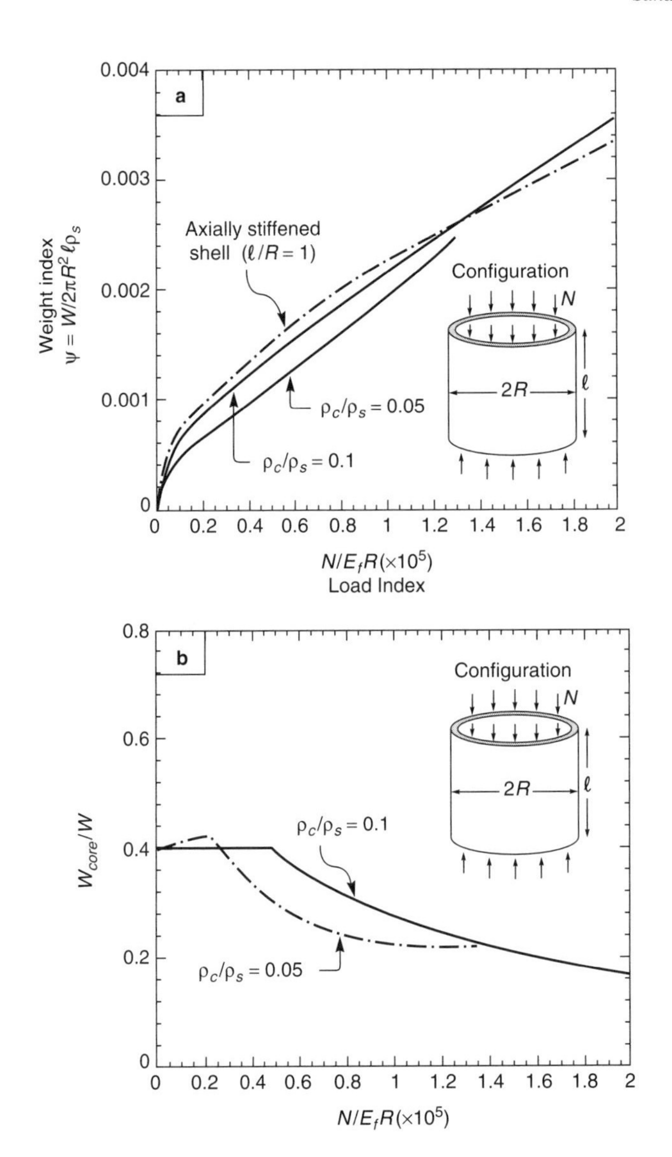

Figure 10.16 *(a) Weight index versus load index for cylindrical sandwich shells. (b) Relative weight versus load index for cylindrical sandwich shells*

design has no relevance. For reference, the result for the optimally designed cylindrical shell with axial hat-stiffeners is repeated from Figure 10.16(a).

This comparison illustrates the weight superiority of foam metal-core sandwich shells over conventional shell construction as well as the potential benefit to be gained by using a core material with the best available stiffness.

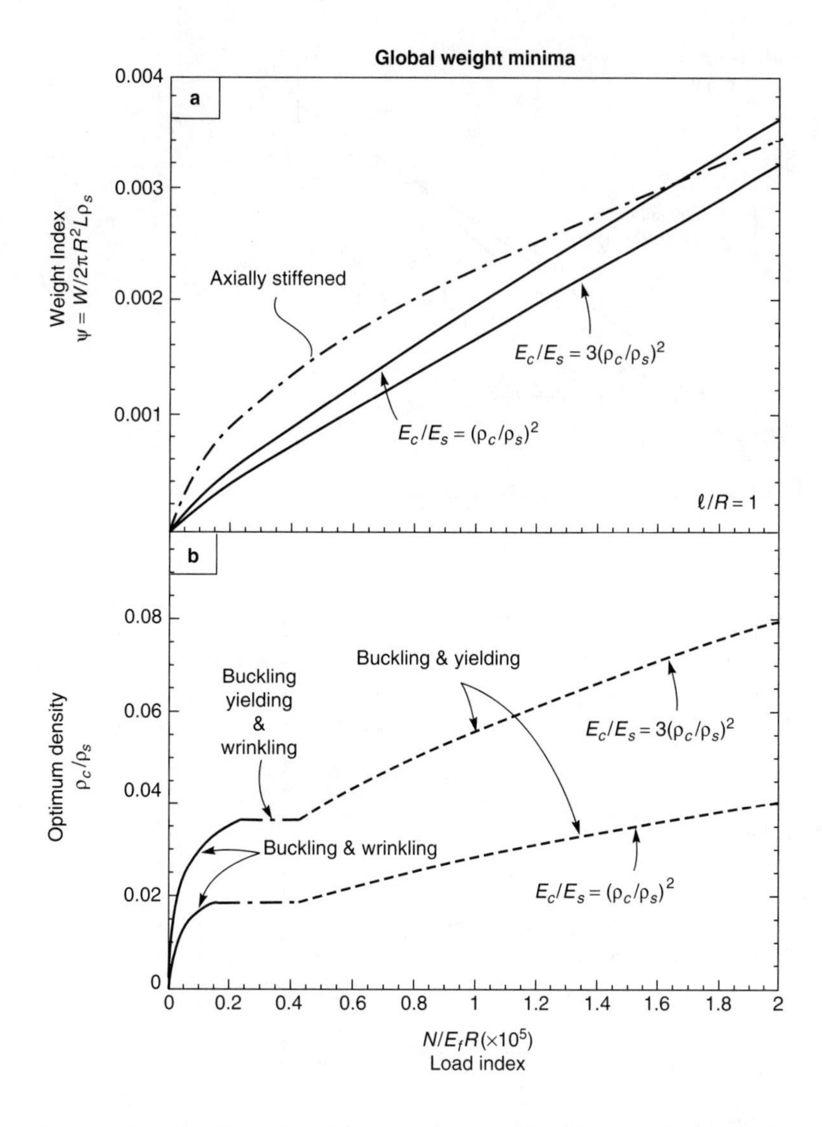

Figure 10.17 *Global weight minima for cylindrical sandwich shells*

Imperfection sensitivity

An important consideration for strength-limited thin-walled construction concerns the influence of imperfections. In most cases, imperfections reduce the buckling loads, sometimes considerably. In shells, imperfections cause out-of-plane bending, which lowers the maximum support load due to two effects: (1) by advancing non-linear collapse and (2) by causing premature plastic yielding, which reduces the local stiffness of the shell and, in turn, hastens collapse. Since they always exist, practical designs take this imperfection sensitivity into account. Generally, experimental results establish a knockdown factor on the theoretical loads that may be used as the design limit with relative impunity. Guidelines for such an experimental protocol are provided by two conclusions from a study of the interaction between plasticity and imperfections in optimally designed axially compressed sandwich shells (Hutchinson and He, 1999). When the perfect cylindrical shell is designed such that buckling and face sheet yielding coincide, buckling in the imperfect shell nearly always occurs prior to plastic yielding. Thus, knockdown factors obtained from standard elastic buckling tests are still applicable to the optimally designed shells. A similar statement pertains when core yielding in the perfect shell is coincident with buckling.

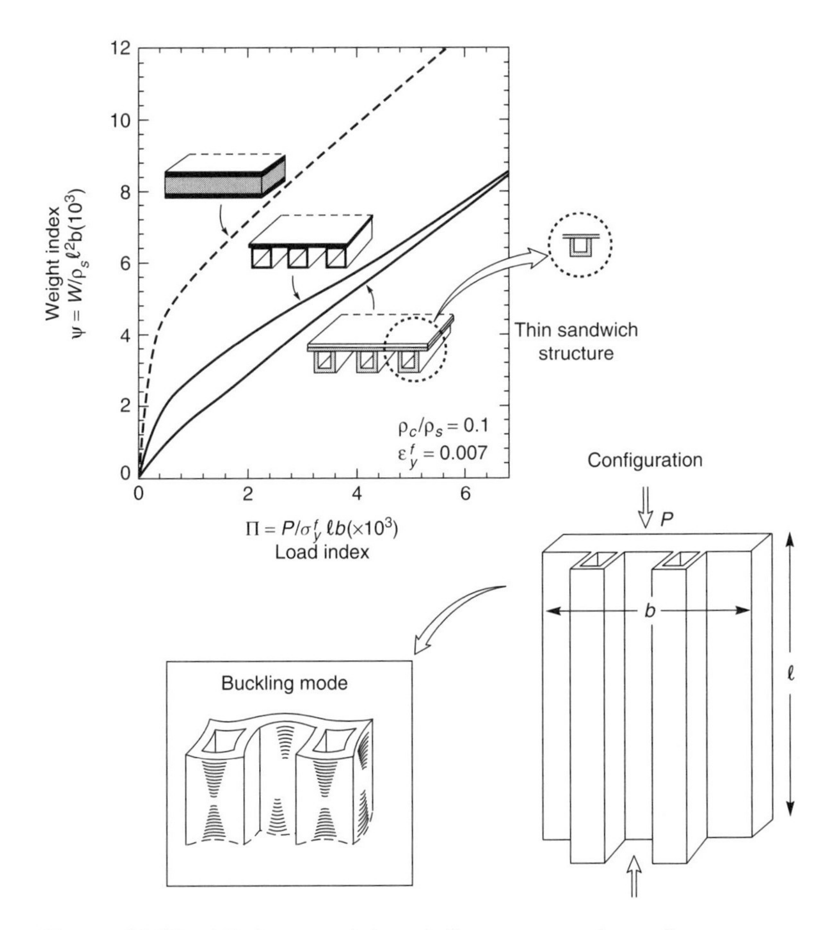

Figure 10.18 *Minimum weight axially compressed panels*

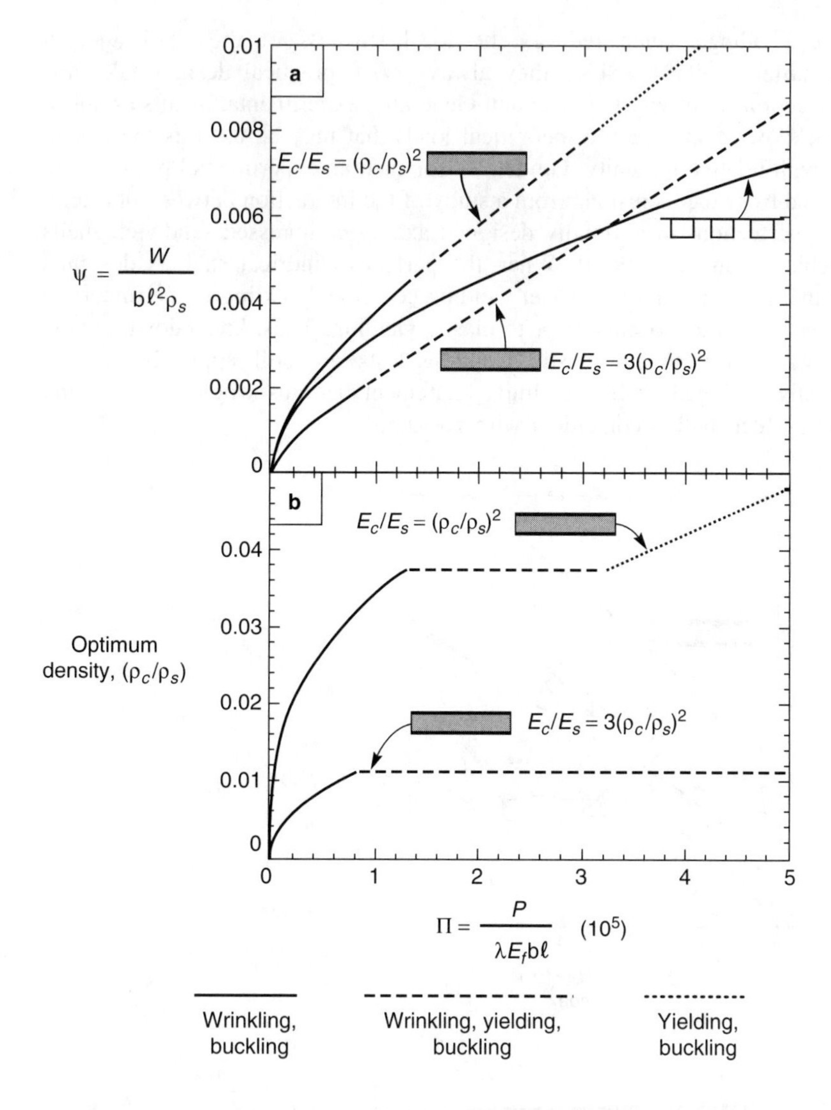

Figure 10.19 *Global weight minima for sandwich and hat-stiffened panels*

Other configurations

Corresponding diagrams for panels and columns (Budiansky, 1999) are presented on Figures 10.18–10.20; associated buckling modes are indicated on the insets. Results for minimum weight flat sandwich panels at a fixed core density, $\rho_c/\rho_s = 0.1$ (Figure 10.18) are not especially promising. There is only a small domain of weight advantage, arising when sandwich construction is used within the stringers, as well as the panels, of a stringer-stiffened configuration. This construction has lowest weight at small levels of load index.

Further minimization with core density leads to more pronounced weight savings (Figure 10.19). In this case, even flat sandwich panels can have lower weight than stringer-stiffened panels, especially at lower levels of load index. The failure modes governing the weight change as the load index changes and the minimum weights coincide with simultaneous occurrences of either two or three modes, as in the case of the optimally designed cylindrical shells. The challenge in taking advantage of the potential weight savings arises in manufacturing and relates to the low relative densities required to realize these performance levels (Figure 10.19) and the need for acceptable morphological quality.

Results for columns (Figure 10.20) indicate that thin-walled sandwich tubes are lighter than foam-filled and conventional tubes, but the beneficial load ranges are small.

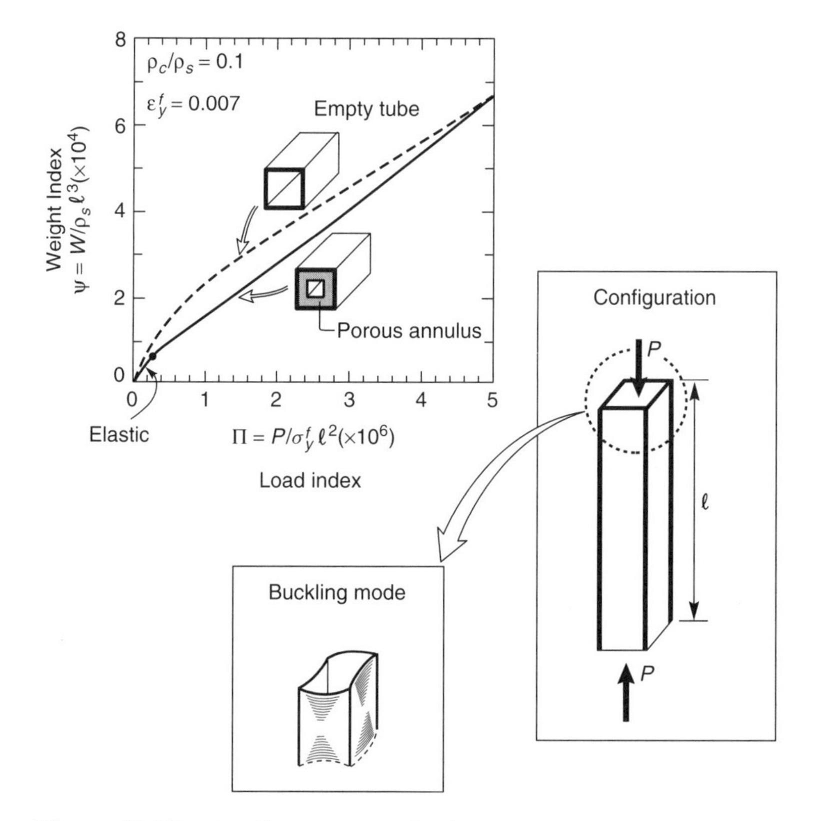

Figure 10.20 *Axially compressed columns*

10.9 Recommendations for sandwich design

For those wishing to explore cellular metal core sandwich construction, the following recommendations are pertinent:

1. Determine the constraints that govern the structure and, in particular, whether it is stiffness or strength-limited.
2. If stiffness-limited, the procedure for determining the minimum weights is straightforward, using the formulae summarized in the tables. It is important to realize that there will always be lighter configurations (especially optimized honeycomb or waffle panels). Those configurations should be explicitly identified, whereupon a manufacturing cost and durability comparison can be made that determines the viability of sandwich construction. Other qualities of the cellular metal may bias the choice. It is important to calculate the domains wherein the weights based on elasticity considerations cannot be realized, because of the incidence of 'inelastic' modes: face yielding, core yielding, face wrinkling. Some help in assessing these limits has been provided.
3. When strength-limited (particularly when buckling-limited), the rules governing sandwich construction are less well formulated. In general, numerical methods are needed to compare and contrast this type of construction with stiffened systems. Some general guidelines are given in this Design Guide; these give insight into the loadings and configurations most likely to benefit from sandwich construction. Configurations unlikely to benefit are also described. It is recommended that where benefits seem likely, detailed simulations and testing should be used to assess the viability of sandwich construction.

References

Allen H.G. (1969) *Analysis and Design of Structural Sandwich Panels*, Pergamon Press, Oxford.

Andrews, E.H., Gioux, G., Onck, P. and Gibson, L.J. (1999) The role of specimen size, specimen shape and surface preparation in mechanical testing of aluminum foams. To appear in *Mat. Sci. and Engineering A*.

Bart-Smith, H. (2000) PhD thesis, Harvard University.

Budiansky, B. (1999) On the minimum weights of compression structures. *Int. J. Solids and Structures* **36**, 3677–3708.

Deshpande, V.S. and Fleck, N.A. (1999) Isotropic constitutive models for metallic foams. To appear in *J. Mech. Phys. Solids*.

Gerard, G. (1956) *Minimum Weight Analysis of Compression Structures*, New York University Press, New York.

Gibson, L.J. and Ashby, M.F. (1997) *Cellular Solids, Structure and Properties*, 2nd edition, Cambridge University Press, Cambridge, Ch. 9, pp. 345 *et seq.*

Harte, A.-M. (1999) Private communication.

Hutchinson, J.W. and He, M.Y. (1999) Buckling of cylindrical sandwich shells with metal foam cores. *Int. J. Solids and Structures* (in press).

Shuaeib, F.M. and Soden, P.D. (1997) Indentation failure of composite sandwich beams. *Composite Science and Technology* **57**, 1249–1259.
Soden, P.D. (1996) Indentation of composite sandwich beams. *J. Strain Analysis* **31**(5), 353–360.
Tennyson, R.C. and Chan, K.C. (1990) Buckling of imperfect sandwich cylinders under axial compression. *Int. J. Solids and Structures* **26**, 1017–1036.
Wilsea, M., Johnson, K.L. and Ashby, M.F. (1975) *Int. J. Mech. Sci.* **17**, 457.

Chapter 11

Energy management: packaging and blast protection

Ideal energy absorbers have a long, flat stress–strain (or load-deflection) curve: the absorber collapses plastically at a constant stress called the plateau stress. Energy absorbers for packaging and protection are chosen so that the plateau stress is just below that which will cause damage to the packaged object; the best choice is then the one which has the longest plateau, and therefore absorbs the most energy. Solid sections do not perform well in this role. Hollow tubes, shells, and metal honeycombs (loaded parallel to the axis of the hexagonal cells) have the right sort of stress–strain curves; so, too, do metal foams.

In crash protection the absorber must absorb the kinetic energy of the moving object without reaching its densification strain ε_D – then the stress it transmits never exceeds the plateau stress. In blast protection, the picture is different. Here it is better to attach a heavy face plate to the absorber on the side exposed to the blast. This is because blast imparts an impulse, conserving momentum, rather than transmitting energy. The calculations become more complicated, but the desired 'ideal absorber' has the same features as those described above.

This chapter reviews energy absorption in metal foams, comparing them with other, competing, systems.

11.1 Introduction: packaging

The function of packaging is to protect the packaged object from damaging acceleration or deceleration. The acceleration or deceleration may be accidental (a drop from a forklift truck, for instance, or head impact in a car accident) or it may be anticipated (the landing-impact of a parachute drop; the launch of a rocket). The damage tolerance of an object is measured by the greatest acceleration or deceleration it can tolerate without harm. Acceleration is measured in units of g, the acceleration due to gravity. Table 11.1 lists typical damage tolerances or 'limiting g-factors' for a range of products.

To protect fully, the package must absorb all the kinetic energy of the object in bringing it to rest. The kinetic energy, W_{KE}, depends on the mass m and the velocity v of the object:

$$W_{KE} = \tfrac{1}{2}mv^2$$

Table 11.1 *Limiting g-factors, a^*, for a number of objects*

Object	*Limiting g-factor, a^**
Human body, sustained acceleration	5–8
Delicate instruments; gyroscopes	15–25
Optical and X-ray equipment	25–40
Computer displays, printers, hard disk drives	40–60
Human head, 36 ms contact time	55–60
Stereos, TV receivers, floppy disk drives	60–85
Household appliances, furniture	85–115
Machine tools, engines, truck and car chassis	115–150

Table 11.2 *Impact velocities for a range of conditions*

Condition	*Velocity (m/s)*
Freefall from forklift truck, drop height 0.3 m	2.4
Freefall from light equipment handler, drop height 0.5 m	3.2
Freefall of carried object or from table, drop height 1 m	4.5
Thrown package, freefall	5.5
Automobile, head impact, roll-over crash in car[a]	6.7
High drag parachute, landing velocity	7
Low drag parachute, landing velocity	13
Automobile, side impact, USA[a]	8.9
Europe[a]	13.8
Automobile, front impact, USA[a]	13.4
Europe[a]	15.6

[a] Current legislation.

Typical velocities for package design are listed in Table 11.2. They lie in the range 2 to 13 m/s (4 to 28 mph). Package design seeks to bring the product, travelling at this velocity, to rest without exceeding its limiting *g*-factor.

11.2 Selecting foams for packaging

Ideal energy absorbers have a long flat stress–strain (or load-deflection) curve like those of Figures 11.1(a) and (b). The absorber collapses plastically at a constant nominal stress, called the plateau stress, σ_{pl}, up to a limiting

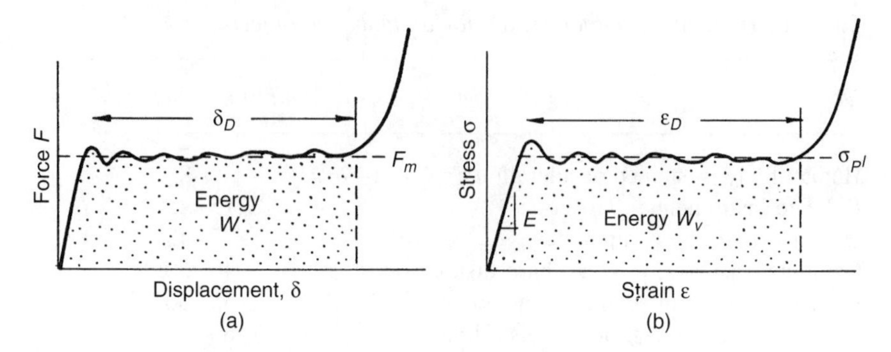

Figure 11.1 *(a) A load-deflection curve and (b) a stress–strain curve for an energy absorber. The area under the flat part ('plateau') of the curves is the useful energy, W, or energy per unit volume, W_v, which can be absorbed. Here F is the force, δ the displacement, σ the stress and ε the strain*

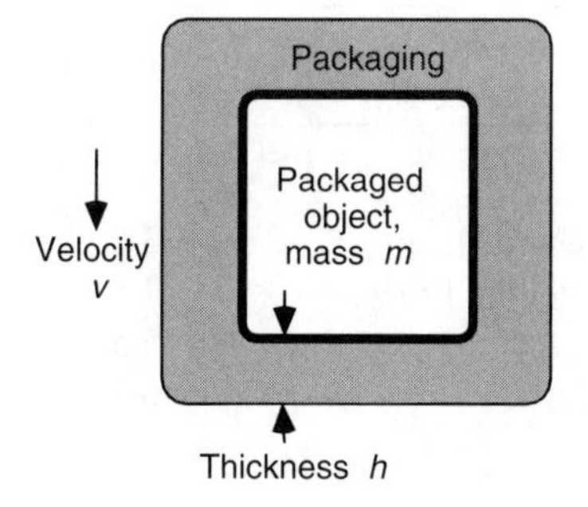

Figure 11.2 *A packaged object. The object is surrounded by a thickness, h, of foam*

nominal strain, ε_D. Energy absorbers for packaging and protection are chosen so that the plateau stress is just below that which will cause damage to the packaged object; the best choice is then the one which has the longest plateau, and therefore absorbs the most energy before reaching ε_D. The area under the curve, roughly $\sigma_{pl}\varepsilon_D$, measures the energy the foam can absorb, per unit initial volume, up to the end of the plateau. Foams which have a stress–strain curve like that shown in Figure 11.1 perform well in this function.

Consider the package shown in Figure 11.2, made from a foam with a plateau stress σ_{pl} and a densification strain ε_D. The packaged object, of mass m, can survive deceleration up to a critical value a^*. From Newton's law the maximum allowable force is

$$F = ma^* \qquad (11.1)$$

If the area of contact between the foam and packaged object is A, the foam will crush when

$$F = \sigma_{pl} A \tag{11.2}$$

Assembling these, we find the foam which will just protect the packaged object from a deceleration a^* is that with a plateau stress

$$\sigma_{pl}^* \leqslant \frac{ma^*}{A} \tag{11.3}$$

The best choice of foam is therefore that with a plateau stress at or below this value which absorbs the most energy.

If packaging of minimum volume is required, we seek the foams that satisfy equation (11.3) and at the same time have the greatest values of the energy per unit volume W_v absorbed by the foam up to densification:

$$W_v = \sigma_{pl} \varepsilon_D \tag{11.4a}$$

If packaging of minimum mass is the goal, we seek foams with the greatest value of energy per unit weight, W_w, absorbed by the foam up to densification:

$$W_w = \frac{\sigma_{pl} \varepsilon_D}{\rho} \tag{11.4b}$$

where ρ is the foam density. And if packaging of minimum cost is sought, we want the foams with the greatest values of energy per unit cost, W_c, absorbed by the foam up to densification:

$$W_c = \frac{\sigma_{pl} \varepsilon_D}{C_m \rho} \tag{11.4c}$$

where C_m is the cost per unit mass of the foam. Figures 11.3, 11.4 and 11.5 show plots of energy per unit volume, mass and cost, plotted against plateau stress, σ_{pl}, for metal foams. These figures guide the choice of foam, as detailed below.

It remains to decide how thick the package must be. The thickness of foam is chosen such that all the kinetic energy of the object is absorbed at the instant when the foam crushes to the end of the plateau. The kinetic energy of the object, $mv^2/2$, must be absorbed by the foam without causing total compaction, when the force rises sharply. Equating the kinetic energy to the energy absorbed by thickness h of foam when crushed to its densification strain ε_D gives

$$\sigma_{pl} \varepsilon_D A h = \tfrac{1}{2} m v^2 \tag{11.5}$$

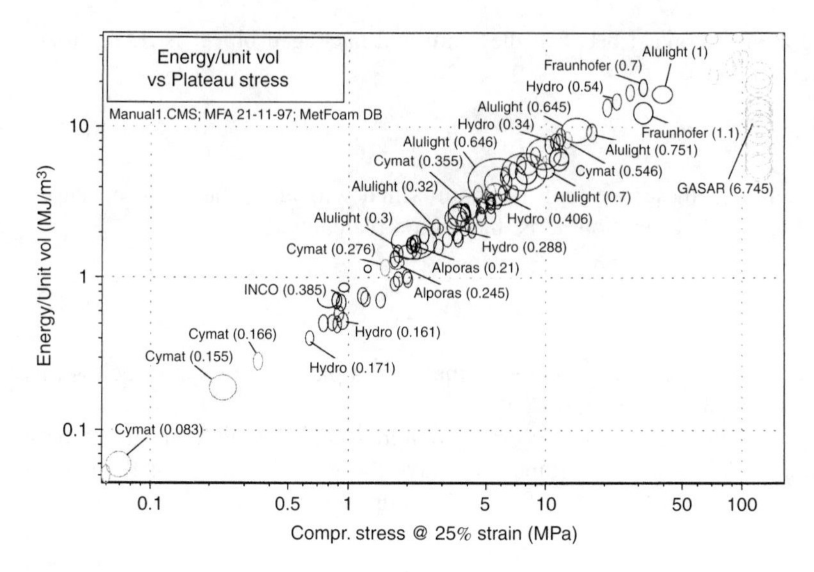

Figure 11.3 *Energy absorbed per unit volume up to densification, plotted against plateau stress (which we take as the compressive strength at 25% strain) for currently available metal foams. Each foam is labeled with its density in Mg/m³*

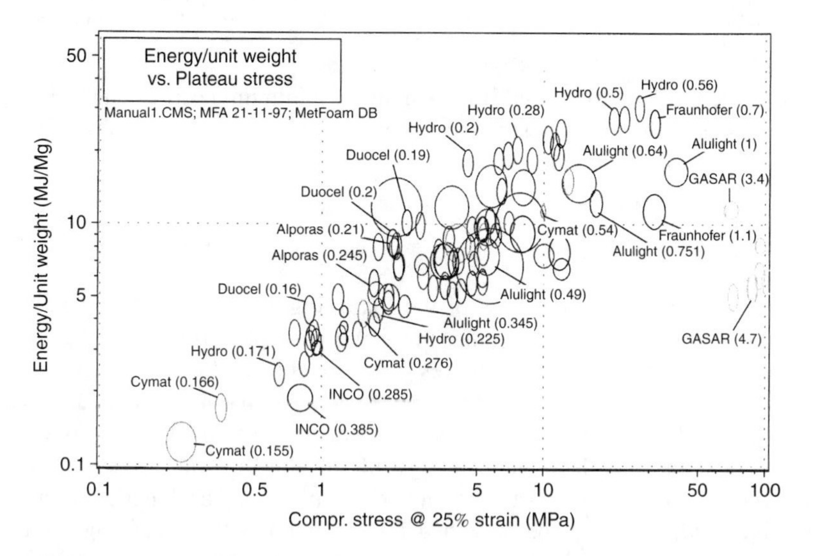

Figure 11.4 *Energy absorbed per unit weight up to densification, plotted against plateau stress (which we take as the compressive strength at 25% strain) for currently available metal foams. Each foam is labeled with its density in Mg/m³*

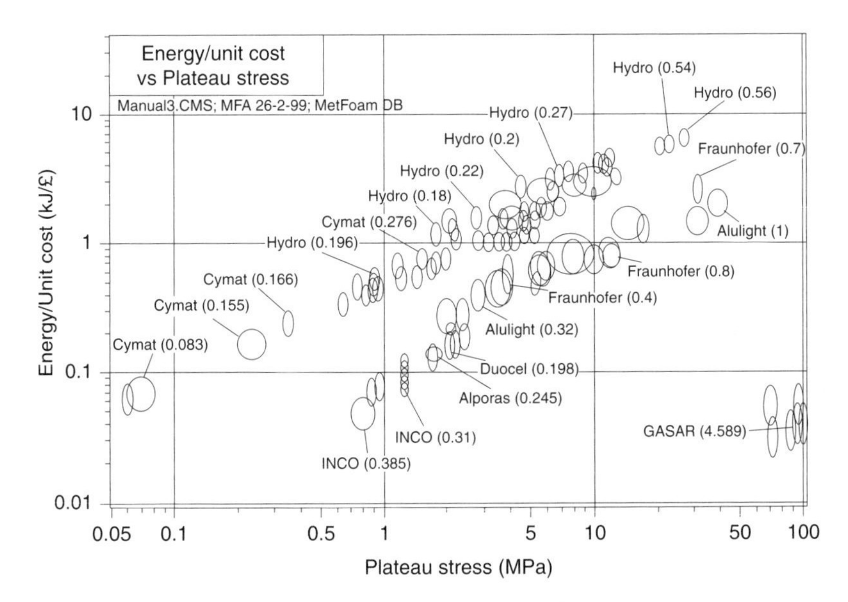

Figure 11.5 *Energy absorbed per unit cost up to densification, plotted against plateau stress (which we take as the compressive strength at 25% strain) for currently available metal foams. Each foam is labeled with its density in Mg/m*^{3}

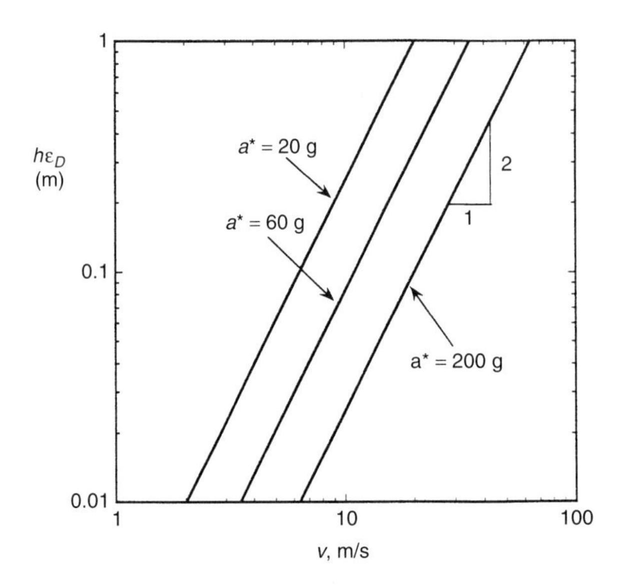

Figure 11.6 *Selection of foam thickness, h, for decelerations of 20 g, 60 g and 200 g*

from which

$$h = \frac{1}{2}\frac{mv^2}{\sigma_{pl}\varepsilon_D A} \tag{11.6}$$

or using equation (11.3):

$$h = \frac{1}{2}\frac{v^2}{a^*\varepsilon_D} \tag{11.7}$$

with $\varepsilon_D = 0.8 - 1.75(\rho/\rho_s)$ (a best fit to recent data for metal foams). Manufacturers' data sheets for the foams give σ_{pl} and ε_D, allowing h to be calculated for a given m, v and a^*. Figure 11.6 shows a plot of equation (11.7). The use of equations (11.6) and (11.7) to design packaging is summarized in Table 11.3.

Table 11.3 *Summary of the steps in initial scoping to select a foam for packaging*

---

(1) Tabulate
- The mass of the product, m
- The limiting g-factor for the product (Table 11.1), a^*
- The impact velocity (Table 11.2), v
- The area of contact between the product and the package, A
- The design objective: minimum volume, or mass, or cost

(2) Calculate the foam crush-stress (the plateau stress) which will just cause the limiting deceleration from equation (11.3).

(3) Plot this as a vertical line on Figures 11.3, 11.4 or 11.5, depending on the objective:
- Minimum volume: Figure 11.3
- Minimum mass: Figure 11.4
- Minimum cost: Figure 11.5

Only foams to the left of the line are candidates; those to the right have plateau stresses which will cause damaging decelerations. Select one or more foams that lie just to the left of the line and as high as possible on the energy scale. Note that the choice depends on the objective.

(4) Use these and the mass and velocity to calculate the required thickness of foam h to absorb all the kinetic energy without reaching the densification strain, using the plateau stress and densification strain which can be read from the data sheet of the chosen foam.

(5) Apply sensible safety factors throughout to allow for margins of error on mass, velocity and foam density.

11.3 Comparison of metal foams with tubular energy absorbers

Thin-walled metal tubes are efficient energy absorbers when crushed axially. By 'efficient' is meant that the energy absorbed per unit of volume or per unit weight is high. How do foams compare with tubes?

When a foam is compressed, its cell walls bend and buckle at almost constant stress until the cell faces impinge. The tube behaves in a different way: it buckles into a series of regular ring-like folds until, when the entire tube has buckled, the fold faces come into contact, as in Figures 11.7 and 11.8 (Andrews *et al.*, 1983; Reid and Reddy, 1983; Wierzbicki and Abramowicz, 1983). The force-displacement curves for both have the approximate shape shown in Figure 11.1(a): a linear-elastic loading line, a long plateau at the constant force, F_m, followed by a steeply rising section as the cell walls or tube folds meet. By dividing F_m by the nominal cross-section of the cylinder (πr^2) and the displacement by the original length (ℓ_0) the force-displacement curve can be converted to an 'effective' stress–strain curve, as shown in Figure 11.1(b). The loading slope is now Young's modulus E (or 'effective' modulus for the tube), the plateau is now at the 'stress' σ_{pl}, and the plateau ends at the densification strain ε_D. The shaded area is the useful energy absorbed per unit volume of structure. Here we compare the energy absorbed by a metal foam with that absorbed by a tube of the same outer dimensions (Seitzberger *et al.*, 1999; Santosa and Wierzbicki, 1998).

Figure 11.9 shows the load-deflection curves for the compression of a foam, a tube and a foam-filled tube. The tubes show regular wave-like oscillations of load, each wave corresponding to the formation of a new fold. A circular tube of length ℓ, outer radius r and wall thickness $t (t \ll r)$ and yield strength

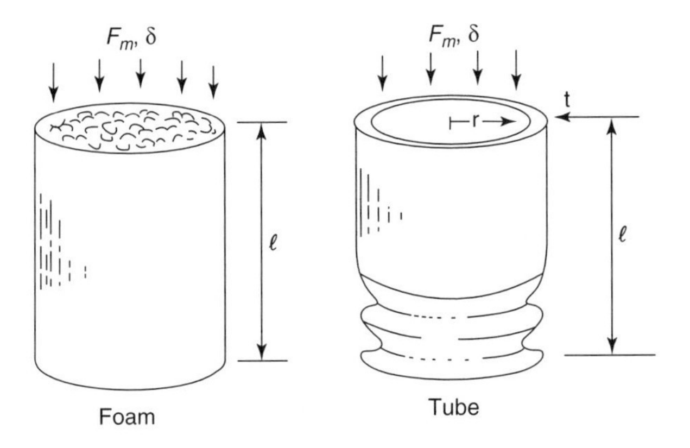

Figure 11.7 *A foam cylinder and a tubular energy absorber*

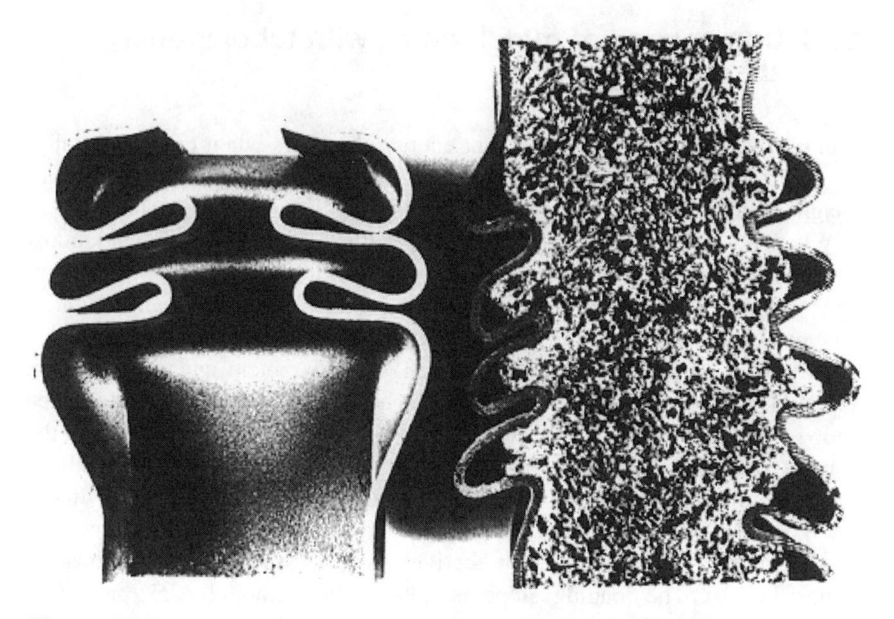

Figure 11.8 *Sections through tubes with and without foam fillings, after partial crushing. (Figure courtesy of Seitzberger et al., 1999)*

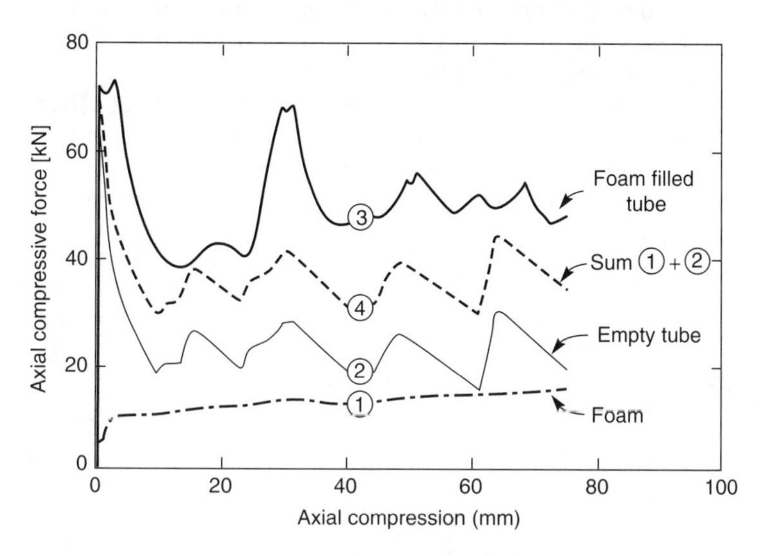

Figure 11.9 *Load-deflection curves for a foam, a tube and a foam-filled tube. The fourth curve is the sum of those for the foam and the tube. The foam-filled tube has a higher collapse load, and can have a higher energy absorption, than those of the sum. (Figure courtesy of Seitzberger et al., 1999)*

σ_{ys} crushes axially at the load

$$F_m = 4\pi r^{1/3} t^{5/3} \sigma_{ys} \tag{11.8}$$

The load remains roughly constant until the folds of the tube lock up at a compaction strain ε_D^{Tube}, giving an axial displacement

$$\delta = \ell \varepsilon_D^{Tube} \tag{11.9}$$

The energy absorbed per unit volume of the tube is then

$$W_v^{Tube} = \frac{F_m \delta}{\pi r^2 \ell} = 4 \left(\frac{t}{r}\right)^{5/3} \sigma_{ys} \varepsilon_D^{Tube}$$

The quantity $2t/r$ is the effective 'relative density' of the tube, ρ/ρ_s, giving

$$W_v^{Tube} = 2^{1/3} \left(\frac{\rho}{\rho_s}\right)^{5/3} \sigma_{ys} \varepsilon_D^{Tube} \tag{11.10}$$

The foam absorbs an energy per unit volume of

$$W_v^{Foam} = C_1 \left(\frac{\rho}{\rho_s}\right)^{3/2} \sigma_{ys} \varepsilon_D^{Foam} \tag{11.11}$$

(using equations (11.4a) and (4.2)) with $C_1 \approx 0.3$. The densification in both the tube and the foam involves the folding of tube or cell walls until they touch and lock-up; to a first approximations the strains ε_D^{Tube} and ε_D^{Foam} are equal at the same relative density. Thus the tube is more efficient than the foam by the approximate factor

$$\frac{W_v^{Tube}}{W_v^{Foam}} \approx 4.2 \left(\frac{\rho}{\rho_s}\right)^{1/6} \tag{11.12}$$

For all realistic values of ρ/ρ_s the tube absorber is more efficient than the foam, on an energy/volume basis, by a factor of about 3.

The equivalent results for energy absorbed per unit weight are

$$W_w^{Tube} = \frac{F_m \delta}{2\pi r t \ell \rho_s} = 2 \left(\frac{t}{r}\right)^{2/3} \frac{\sigma_{ys}}{\rho_s} \varepsilon_D^{Tube}$$

or, replacing $2t/r$ by ρ/ρ_s,

$$W_w^{Tube} = 2^{1/3} \left(\frac{\rho}{\rho_s}\right)^{2/3} \frac{\sigma_{ys}}{\rho_s} \varepsilon_D^{Tube} \tag{11.13}$$

That for the foam is

$$W_w^{Foam} = C_1 \left(\frac{\rho}{\rho_s}\right)^{1/2} \frac{\sigma_{ys}}{\rho_s} \varepsilon_D^{Foam} \tag{11.14}$$

giving the same ratio as before – equation (11.12) – and with the same conclusions.

More detailed calculations and measurements bear out the conclusions reached here. Figure 11.10 shows the calculated energy per unit mass absorbed by tubes plotted against the upper-bound collapse stress (the plateau stress) compared with measured values for foams. Axially compressed tubes outperform foams by a small but significant margin. Foams retain the advantage that they are isotropic, absorbing energy equally well for any direction of impact. Tubes hit obliquely are less good.

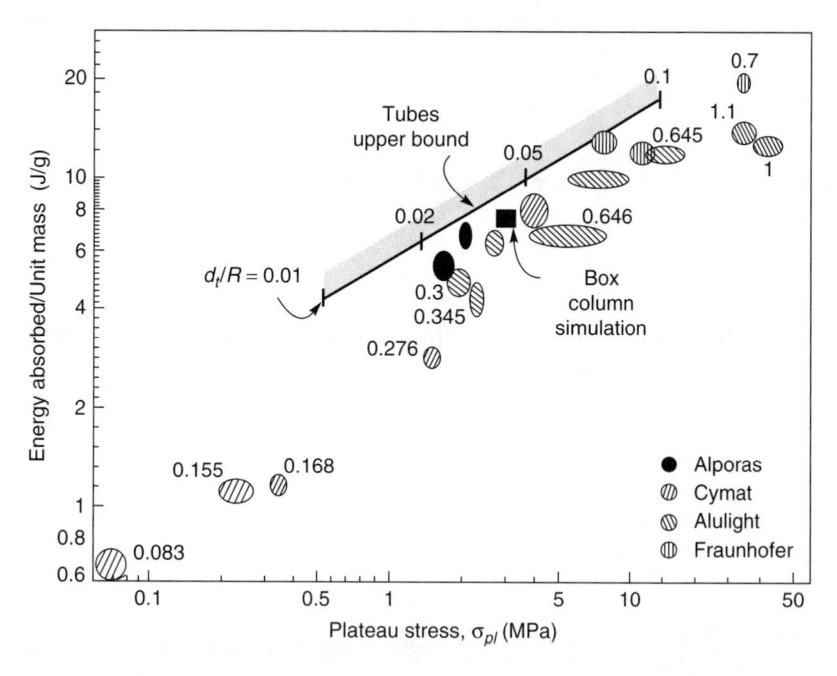

Figure 11.10 *The energy absorbed per unit mass by tubes (full line) and by metal foams, plotted against plateau stress, σ_{pl}. The data for tubes derive from an upper bound calculation of the collapse stress. Each foam is labeled with its density in Mg/m^3*

Foam-filled sections

A gain in efficiency is made possible by filling tubes with metal foam. The effect is demonstrated in Figure 11.9 in which the sum of the individual loads carried by a tube and a foam, at a given displacement, is compared with the measured result when the foam is inserted in the tube. This synergistic enhancement is described by

$$W_v^{Filled\ tube} = W_v^{Tube} + W_v^{Foam} + W_v^{Int.} \tag{11.15}$$

where the additional energy absorbed, $W_v^{Int.}$, arises from the interaction between the tube and the foam. This is because the foam provides internal support for the tube wall, shortening the wavelength of the buckles and thus creating more plastic folds per unit length (Abramowicz and Wierzbicki, 1988; Hanssen *et al.*, 1999) A similar gain in energy-absorbing efficiency is found in the bending of filled tubes (Santosa *et al.*, 1999).

The presence of the foam within the tube reduces the stroke δ before the folds in the tube lock up, but, provided the density of the foam is properly chosen, the increase in the collapse load, F_m, is such that the energy $F_m\delta$ increases by up to 30% (Seitzberger *et al.*, 1999).

11.4 Effect of strain rate on plateau stress

Impact velocities above about 1 m/s (3.6 km/h) lead to strain rates which can be large: a 10 m/s impact on a 100 mm absorber gives a nominal strain rate of 100/s. It is then important to ask whether the foam properties shown here and in Figures 4.6–4.11 of Section 4, based on measurements made at low strain rates (typically 10^{-2}/s), are still relevant.

Tests on aluminum-based foams show that the dependence of plateau stress on strain rate is not strong (Kenny, 1996; Lankford and Danneman, 1998; Deshpande and Fleck, 2000). Data are shown in Figures 11.11 and 11.12 for an Alporas closed-cell foam and an ERG Duocel (Al-6101-T6) open-cell foam. They suggests that the plateau stress, σ_{pl}, increases with strain rate $\dot{\varepsilon}$ by, at most, 30%, over the range

$$3.6 \times 10^{-3}/s < \dot{\varepsilon} < 3.6 \times 10^{+3}/s$$

Tests on magnesium-based foams (Mukai *et al.*, 1999) show a stronger effect. For these, the plateau stress is found to increase by roughly a factor of 2 over the same range of strain rate.

It is important to separate the effect of strain rate and impact velocity on the dynamic response of a metallic foam. The negligible effect of strain rate is associated with the fact that aluminum displays only a minor strain-rate sensitivity. In contrast, material inertia leads to enhanced stresses at high impact velocities. At the simplest level, the effects can be understood in terms of a one-dimensional shock wave analysis, elaborated below.

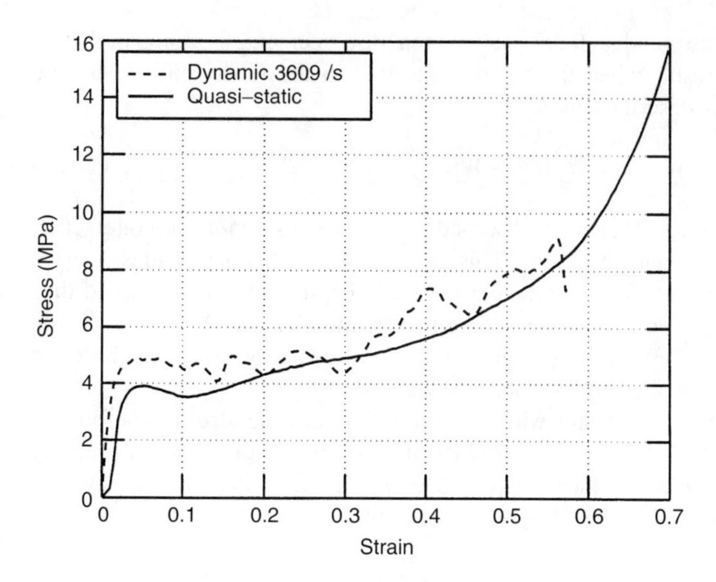

Figure 11.11 *Stress–strain curves for Alulight foam with a relative density of 0.18 at two strain rates: 3.6 × 10^{-3}/s and 3.6 × 10^{+3}/s (Deshpande and Fleck, 2000)*

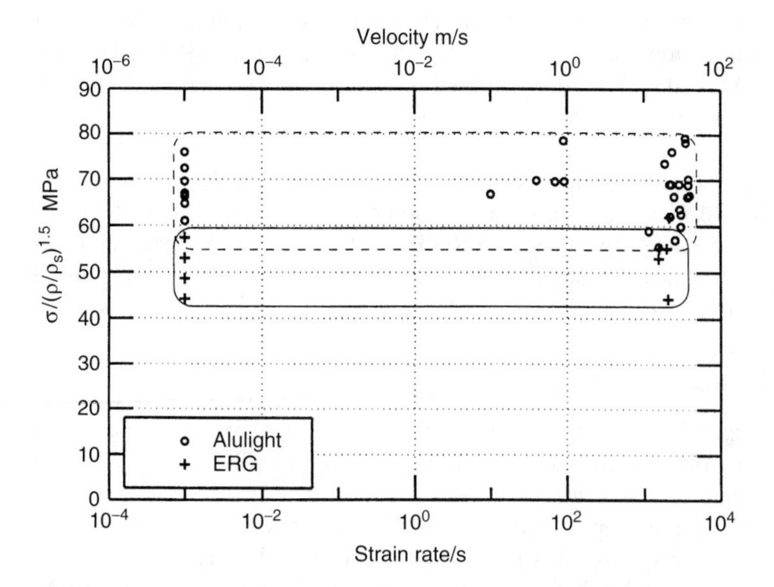

Figure 11.12 *The plateau stress corrected for relative density, plotted against strain rate. It is essentially independent of strain rate up to 3.6 × 10^{+3}/s (Deshpande and Fleck, 2000)*

11.5 Propagation of shock waves in metal foams

When a metal foam is impacted at a sufficiently high velocity, made precise below, a plastic shock wave passes through it and the plateau stress rises. Consider the idealized nominal compressive stress–strain curve for a metallic foam, shown in Figure 11.13. It has an initial elastic modulus, E, and a plateau stress, σ_{pl}, before compaction occurs at a nominal densification strain, ε_D. When such a foam is impacted an elastic wave propagates through it; and if the stress rises above σ_{pl}, this is followed by a plastic shock wave. In the simplified one-dimensional case sketched in Figure 11.14(a) it is imagined that the bar is initially stationary and stress-free. At a time $t = 0$, the left-hand end of the bar is subjected to a constant velocity, V. In response, an elastic wave travels

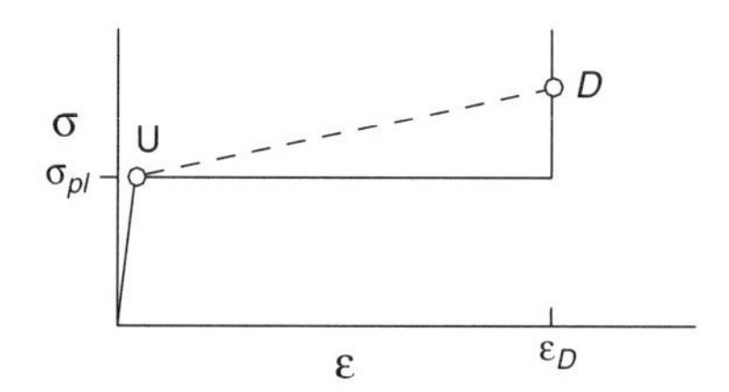

Figure 11.13 *A schematic compressive stress–strain curve for a metal foam; the stress jumps from the plateau level σ_{pl} at U to the value σ_D at D*

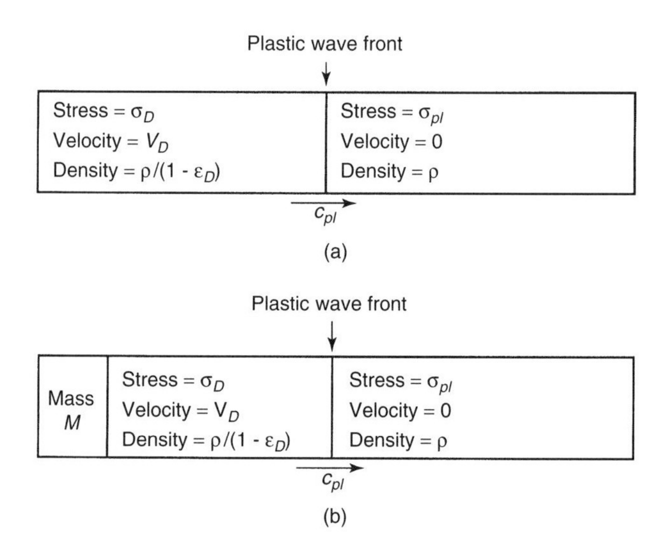

Figure 11.14 *(a) The stress, velocity on either side of the plastic shock-wave front for an impacted foam (b) The stress, velocity on either side of the plastic shock wave front for an impacted foam carrying a buffer plate of mass M*

quickly along the bar at an elastic wave speed $c_{el} = \sqrt{E/\rho}$ and brings the bar to a uniform stress of σ_{pl} and to a negligibly small velocity of $v = \sigma_{pl}/(\rho c_{el})$. Trailing behind this elastic wave is the more major disturbance of the plastic shock wave, travelling at a wave speed c_{pl}. Upstream of the plastic shock front the stress is σ_{pl}, and the velocity is $v_U \approx 0$. Downstream of the shock the stress and strain state is given by the point D on the stress–strain curve: the compressive stress is σ_D, the strain equals the densification strain, ε_D, and the foam density has increased to $\rho_D = \rho/(1 - \varepsilon_D)$.

Momentum conservation for the plastic shock wave dictates that the stress jump $(\sigma_D - \sigma_{pl})$ across the shock is related to the velocity jump, v_D, by

$$(\sigma_D - \sigma_{pl}) = \rho c_{pl} v_D \tag{11.16}$$

and material continuity implies that the velocity jump, v_D, is related to the strain jump, ε_D, by

$$v_D = c_{pl}\varepsilon_D \tag{11.17}$$

Elimination of v_D from the above two relations gives the *plastic wave speed*, c_{pl}:

$$c_{pl} = \sqrt{\frac{(\sigma_D - \sigma_{pl})}{\rho\varepsilon_D}} \tag{11.18}$$

This has the form

$$c_{pl} = \sqrt{\frac{E_t}{\rho}} \tag{11.19}$$

where the tangent modulus, E_t, is the slope of the dotted line joining the downstream state to the upstream state (see Figure 11.13), defined by

$$E_t = \frac{(\sigma_D - \sigma_{pl})}{\varepsilon_D} \tag{11.20}$$

The location of the point D on the stress–strain curve depends upon the problem in hand. For the case considered above, the downstream velocity, v_D, is held fixed at the impact velocity, V; then, the wave speed c_{pl} is $c_{pl} \equiv v_D/\varepsilon_D = V/\varepsilon_D$, and the downstream stress σ_D is constant at $\sigma_D \equiv \sigma_{pl} + \rho c_{pl} v_D = \sigma_{pl} + \rho V^2/\varepsilon_D$. This equation reveals that the downstream stress, σ_D, is the sum of the plastic strength of the foam, σ_{pl}, and the hydrodynamic term $\rho V^2/\varepsilon_D$. A simple criterion for the onset of inertial loading effects in foams is derived by defining a transition speed V_t for which the hydrodynamic contribution to strength is 10% of the static contribution, giving

$$V_t = \sqrt{\frac{0.1\sigma_{pl}\varepsilon_D}{\rho}} \tag{11.21}$$

Recall that the plateau strength for metal foams is approximated by (Table 4.2)

$$\frac{\sigma_{pl}}{\sigma_{ys}} = C_1 \left(\frac{\rho}{\rho_s}\right)^{3/2} \tag{11.22}$$

where σ_{ys} is the yield strength of the solid of which the foam is made and C_1 is a constant with a value often between 0.2 and 0.3. The densification strain scales with relative density according to $\varepsilon_D = \alpha - \beta(\rho/\rho_s)$, where $\alpha \approx 0.8$ and $\beta \approx 1.75$. Hence, the transition speed, V_t, depends upon foam density according to

$$V_t = \left(0.1C_1\frac{\sigma_{ys}}{\rho_s}\right)^{1/2}\left(\frac{\rho}{\rho_s}\right)^{1/4}\left(\alpha - \beta\frac{\rho}{\rho_s}\right)^{1/2} \tag{11.23}$$

Examination of this relation suggests that V_t shows a maximum at $\rho/\rho_s = \alpha/3\beta = 0.15$. Now insert some typical values. On taking $C_1 = 0.3$, $\rho/\rho_s = 0.15$, $\sigma_{ys} = 200\,\text{MPa}$ and $\rho_s = 2700\,\text{kgm}^{-3}$, we find $V_t = 21.5\,\text{ms}^{-1}$ ($77\,\text{km h}^{-1}$). For most practical applications in ground transportation, the anticipated impact speeds are much less than this value, and we conclude that the quasi-static strength suffices at the conceptual design stage.

Kinetic energy absorber

Insight into the optimal design of a foam energy absorber is gained by considering the one-dimensional problem of end-on impact of a long bar of foam of cross-sectional area A by a body of mass M with an impact velocity V_0, as sketched in Figure 11.14(b). After impact, a plastic shock wave moves from the impact end of the bar at a wave speed c_{pl}. Consider the state of stress in the foam after the plastic wave has travelled a distance ℓ from the impacted end. Upstream of the shock, the foam is stationary (except for a small speed due to elastic wave effects) and is subjected to the plateau stress σ_{pl}. Downstream, the foam has compacted to a strain of ε_D, is subjected to a stress σ_D and moves at a velocity v_D equal to that of the mass M. An energy balance gives

$$\frac{1}{2}\left(M + \frac{\rho}{1-\varepsilon_D}A\ell\right)v_D^2 + \sigma_{pl}\varepsilon_D A\frac{\ell}{1-\varepsilon_D} = \frac{1}{2}MV_0^2 \tag{11.24}$$

Using the relation $v_D = c_{pl}\varepsilon_D$ and the momentum balance $\sigma_D = \sigma_{pl} + \rho c_{pl} v_D$, the downstream compressive stress, σ_D, exerted on the impacting mass is

$$\sigma_D = \sigma_{pl} + \frac{\rho}{\varepsilon_D} \frac{MV_0^2 - 2\sigma_{pl} A\ell\varepsilon_D/(1-\varepsilon_D)}{M + \rho A\ell/(1-\varepsilon_D)} \tag{11.25}$$

Thus, the compressive stress decelerating the mass M decreases with the length of foam compacted, $\ell/(1-\varepsilon_D)$. In the limit $\ell = 0$ the peak compressive stress on the mass is

$$(\sigma_D)_{peak} = \sigma_{pl} + \frac{\rho V_0^2}{\varepsilon_D} \tag{11.26}$$

in agreement with the findings of the shock wave analysis above. The length of foam $\ell/(1-\varepsilon_D)$ required to arrest the mass is determined by putting $v_D = 0$ in the above equation, giving

$$\frac{\ell}{1-\varepsilon_D} = \frac{MV_0^2}{2\sigma_{pl}\varepsilon_D A} \tag{11.27}$$

On noting that $\sigma_{pl}\varepsilon_D$ equals the energy, W, absorbed by the foam per unit volume, we see that the minimum length of foam required for energy absorption is obtained by selecting a foam with a maximum value of W, consistent with a value of upstream stress σ_{pl} which does not overload the structure to which the foam is attached. The plots of W versus σ_{pl} for metallic foams (Figures 11.3, 11.4 and 11.5) are useful for this selection process.

11.6 Blast and projectile protection

Explosives create a pressure wave of approximately triangular profile, known as a 'blast' (Smith and Hetherington, 1994). Protection against blast involves new features. The blast imparts an impulse, J_i, per unit area of a structure, equal to the integral of the pressure over time:

$$J_i = \int p\mathrm{d}t \tag{11.28}$$

The blast wave is reflected by a rigid structure, and the details of the pressure–time transient depend on the orientation of the structure with respect to the pressure wave. In design, it is conservative to assume that the structure is at normal incidence and fully reflects the blast. Figure 11.15 shows the peak pressure, p_0, and the resulting impulse, J_i, caused by the detonation of a charge of TNT, at a radial distance, R, from the charge. Curves are shown for a reflected blast in air, and in water. The impulse and the distance are normalized by the cube root of the mass, M, of the charge in kg. As an

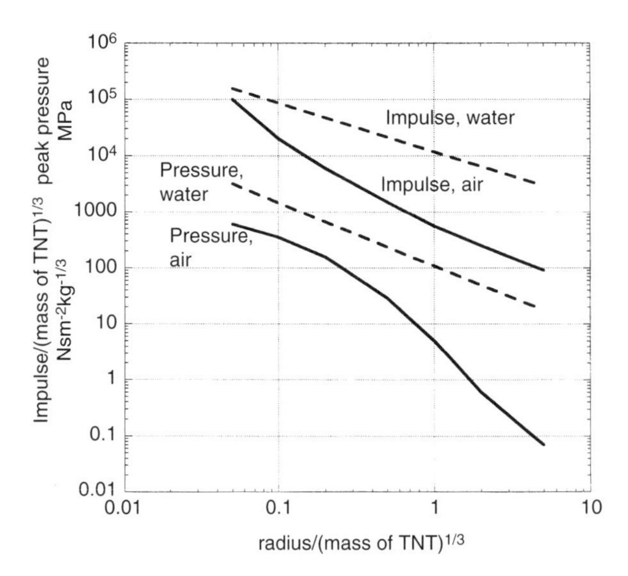

Figure 11.15 *Peak pressure and impulse as a function of distance R from an explosion of a mass M of TNT*

example, a charge of 1 kg of TNT in water produces a peak pressure of about 100 MPa and an impulse of 10^4Ns/m^2 at a distance of 1 m. The curves for water blast are approximated by the formulae

$$p_0 = 108 \left(\frac{M^{1/3}}{R}\right)^{1.13} \quad \text{MPa} \tag{11.29}$$

and

$$J_i = 1.185 \times 10^4 M^{1/3} \left(\frac{M^{1/3}}{R}\right)^{0.86} \quad \text{Ns m}^{-2} \tag{11.30}$$

where the mass, M, of TNT is given in kilograms and the distance from the explosion, R, is given in metres. The energy content of other common chemical explosives is similar to that of TNT, as shown in Table 11.4. In order to estimate the blast from other explosives the simplest method is to scale the mass of the explosive by its energy content relative to that of TNT: this scale factor is included in the table.

Blast protection (and protection from projectile impact, which is treated in a similar way) is achieved by attaching a heavy buffer plate, mounted on an energy absorber, to the face of the object to be protected. The impulse accelerates the buffer plate; its kinetic energy is dissipated in a benign way by the energy absorber. Let the buffer plate have a thickness b and density ρ_b. Then

Table 11.4 *Energy density and TNT equivalents of explosives*

Explosive	*Mass specific energy Q_x (KJ/kg)*	*TNT equivalent (Q_x/Q_{TNT})*
Amatol 80/20 (80% ammonium nitrate, 20% TNT)	2650	0.586
Compound B (60% RDX, 40% TNT)	5190	1.148
RDX (Cyclonite)	5360	1.185
HMX	5680	1.256
Lead azide	1540	0.340
Mercury fulminate	1790	0.395
Nitroglycerine (liquid)	6700	1.481
PETN	5800	1.282
Pentolite 50/50 (50% PETN, 50% TNT)	5110	1.129
Tetryl	4520	1.000
TNT	4520	1.000
Torpex (42% RDX, 40% TNT, 18% aluminum)	7540	1.667
Blasting gelatin (91% nitroglycerine, 7.9% nitrocellulose, 0.9% antacid, 0.2% water)	4520	1.000
60% Nitroglycerine dynamite	2710	0.600

the impulse, J_i, imparts a momentum, M_i, to a unit area of the face plate where

$$M_i = \rho_b b v = J_i \tag{11.31}$$

and thereby accelerates the plate to a velocity v. At this point it has a kinetic energy

$$U_i = \frac{1}{2}\rho_b b v^2 = \frac{J_i^2}{2\rho_b b} \tag{11.32}$$

and it is this that the energy absorber must dissipate. Note that the thicker and heavier the buffer plate, the lower is the kinetic energy that the absorber must dissipate.

The selection of a metal foam as an energy absorber follows the method of Section 11.2. It is necessary to absorb U_i per unit area at a plateau stress σ_{pl} which will not damage the protected object. Let the energy absorbed per unit volume up to densification by a foam with a plateau stress σ_{pl} be W_{vol}. Then the thickness h_{blast} of foam required to absorb the blast is

$$h_{blast} = \frac{J_i^2}{2\rho_b b W_{vol}} \tag{11.33}$$

The efficiency of absorption is maximized by using a heavy buffer plate ($\rho_b b$) and choosing the foam with the greatest W_{vol} for a given σ_{pl}.

An alternative strategy might be to minimize the combined mass per unit area, m_t of the buffer plate and the foam. The mass per unit area of the buffer plate is $m_b = \rho_b b$, and the mass per unit area of the foam is $m_f = \rho\, h_{blast}$. Substitution for h_{blast} from equation (11.33) into the expression for m_f gives

$$m_t = m_b + m_f = \rho_b b + \frac{\rho J_i^2}{2\rho_b b W_{vol}}$$

and minimization of m_t with respect to the buffer plate mass $m_b = \rho_b b$ gives $m_b = m_f$, and

$$m_t = 2m_f = J_i\sqrt{\frac{2\rho}{W_{vol}}}$$

The thickness of the buffer plate is related to that of the foam simply by $b = \rho\, h_{blast}/\rho_b$.

An example

A charge of 1 kg of TNT in air produces a pressure pulse $p = 5$ MPa (50 atmospheres), generating an impulse $J_i = 600$ Ns/m^2 at a distance 1m from the charge. A steel buffer plate ($\rho_b = 7900$ kg/m^3) 5 mm thick acquires a velocity $v = 15.2$ m/s and a kinetic energy $U_i = 4.6$ kJ/m^2. The structure can support a pressure of 0.3 MPa (3 atmospheres). The selection chart of Figure 11.3 indicates that a Cymat aluminum foam of density 0.155 Mg/m^3 has a plateau stress just below this, and absorbs $W_{vol} = 200$ kJ/m^3. From equation (11.33) the required thickness of foam is 25 mm.

References

Abramowicz, W. and Wierzbicki, T. (1988) Axial crushing of foam-filled columns. *Int. J. Mech. Sci.* **30**(3/4), 263–271.

Andrews, K.R.F., England, G.L. and Ghani, E. (1983) Classification of the axial collapse of cylindrical tubes. *Int. J. Mech. Sci.* **25**, 687–696

Deshpande, V. and Fleck, N.A. (2000) High strain rate compressive behavior of aluminum alloy foams. *Int. J. Impact Engng.* **24**, 277–298.

Hanssen, A.G., Langseth, M. and Hopperstad, O.S. (1999) Static crushing of square aluminum extrusions with aluminum foam filler. *Int. J. Mech. Sci.* **41**, 967–993.

Kenny, L.D. (1996) Mechanical properties of particle stabilised aluminum foams. *Materials Science Forum* **217–222**, 1883–1890.

Lankford, J. and Danneman, K. (1998) in Shwartz, D.S., Shih, D.S., Evans, A.G. and Wadley, H.N.G. (eds), *Porous and Cellular Materials for Structural Application*, Materials Research Society Proceedings, Vol. 521, MRS, Warrendale, PA, USA.

Mukai, T., Kanahashi, H., Higashi, K., Yamada, Y., Shimojima, K., Mabuchi, M., Miyoshi, T. and Tieh, T.G. (1999) Energy absorption of light-weight metallic foams under dynamic loading. In Banhart, J., Ashby, M.F. and Fleck, N.A. (eds), *Metal Foams and Foam Metal Structures*, Proc. Int. Conf. *Metfoam'99*, 14–16 June 1999, MIT Verlag, Bremen, Germany.

Reid, S.R. and Reddy, T.Y. (1983) *Int. J. Impact Engng.* **1**, 85–106.

Santosa, S. and Wierzbicki, T. (1998) Crash behavior of box columns filled with aluminum honeycomb or foam. *Computers and Structures* **68**(4), 343–367.

Santosa, S., Banhart, J. and Wierzbicki, T. (1999) Bending crush behavior of foam-filled sections. In Banhart, J., Ashby, M.F. and Fleck, N.A. (eds), *Metal Foams and Foam Metal Structures*, Proc. Int. Conf. *Metfoam'99*, 14–16 June 1999, MIT Verlag, Bremen, Germany.

Seitzberger, M., Rammerstorfer, F.G., Gradinger, R., Degisher, H.P., Blaimschein, M. and Walch, C. (1999) Axial crushing of foam-filled columns. To appear in *Int. J. Sol. Struct.*

Seitzberger, M., Rammerstorfer, F.G., Degisher, H.P. and Gradinger, R. (1997) Crushing of axially compressed steel tubes filled with aluminum foam. *Acta Mechanica* **125**, 93–105.

Smith, P.D. and Hetherington, J.G. (1994) *Blast and Ballistic Loading of Structures*, Butterworth-Heinemann, Oxford.

Wierzbicki, T. and Abramowicz, W. (1983) On the crushing mechanics of thin-walled structures. *Int. J. Solids Structures* **29**, 3269–3288.

Chapter 12

Sound absorption and vibration suppression

The ability to damp vibration coupled with mechanical stiffness and strength at low weight makes an attractive combination. Automobile floors and bulkheads are examples of structures the primary function of which is to carry loads, but if this is combined with vibration damping and sound absorption the product quality is enhanced.

Metal foams have higher mechanical damping than the solid of which they are made, but this is not the same as sound absorption. Sound absorption means an incident sound wave is neither reflected nor transmitted; its energy is absorbed in the material. There are many ways in which this can happen: by direct mechanical damping in the material itself, by thermo-elastic damping, by viscous losses as the pressure wave pumps air in and out of cavities in the absorber and by vortex-shedding from sharp edges. Sound is measured in decibels, and this is a logarithmic measure, in accord with the response of the ear. The result is that – as far as human perception is concerned – a sound-absorption coefficient in the acoustic range of 0.5 (meaning that half the incident energy is absorbed) is not much good. To be really effective, the absorption coefficient must be exceed 0.9. The best acoustic absorbers easily achieve this.

Are metal foams good sound absorbers? Data described in this chapter suggest an absorption coefficient of up to 0.85 – good, but not as good as materials such as felt or fiberglass. More significant is that the high flexural stiffness and low mass of foam and foam-cored panels result in high natural vibration frequencies, and this makes them hard to excite. So while metal foams and metfoam-cored panels offer some potential for vibration and acoustic management, their greater attraction lies in the combination of this attribute with others such as of stiffness at light weight, mechanical isolation, fire protection and chemical stability.

12.1 Background: sound absorption in structural materials

Sound is caused by vibration in an elastic medium. In air at sea level it travels at a velocity of 343 m/s, but in solids it travels much faster: in both steel and aluminum the sound velocity is about 5000 m/s. The wave velocity, v, is related to wavelength λ_s and frequency f by $v = \lambda_s f$. To give a perspective: the (youthful) human ear responds to frequencies from about 20 to about 20 000 Hz, corresponding to wavelengths of 17 m to 17 mm. The bottom note on a piano is 28 Hz; the top note 4186 Hz. The most important range, from an acoustic point of view, is roughly 500–4000 Hz.

Sound pressure is measured in Pascals (Pa), but because audible sound pressure has a range of about 10^6, it is more convenient to use a logarithmic scale with units of decibels (dB). The decibel scale compares two sounds and therefore is not absolute. Confusingly, there are two decibel scales in use (Beranek, 1960). The decibel scale for sound pressure level (SPL) is defined as

$$\mathrm{SPL} = 10\log_{10}\left(\frac{p_{rms}}{p_0}\right)^2 = 20\log_{10}\left(\frac{p_{rms}}{p_0}\right) \tag{12.1a}$$

where p_{rms} is the (mean square) sound pressure and p_0 is a reference pressure, taken as the threshold of hearing (a sound pressure of 20×10^{-6} Pa). The decibel scale for sound power level (PWL) is defined by

$$\mathrm{PWL} = 10\log_{10}\left(\frac{W}{W_0}\right) \tag{12.1b}$$

where W is the power level and W_0 is a reference power ($W_0 = 10^{-12}$ watt if the metric system is used, 10^{-13} watt if the English system is used). The two decibel scales are closely related, since sound power is proportional to p_{rms}^2. In practice it is common to use the SPL scale. Table 12.1 shows sound levels, measured in dB using this scale.

Table 12.1 *Sound levels in decibels*

Threshold of hearing	0
Background noise in quiet office	50
Road traffic	80
Discotheque	100
Pneumatic drill at 1 m	110
Jet take-off at 100 m	120

The sound-absorption coefficient measures the fraction of the energy of a plane sound wave which is absorbed when it is incident on a material. A material with a coefficient of 0.9 absorbs 90% of the sound energy, and this corresponds to a change of sound level of 10 dB. Table 12.2 shows sound-absorption coefficients for a number of building materials (Cowan and Smith, 1988)

Table 12.2 *Sound-absorption coefficient at indicated frequency*

Material	*500 Hz*	*1000 Hz*	*2000 Hz*	*4000 Hz*
Glazed tiles	0.01	0.01	0.02	0.02
Concrete with roughened surface	0.02	0.03	0.04	0.04
Timber floor on timber joists	0.15	0.10	0.10	0.08
Cork tiles on solid backing	0.20	0.55	0.60	0.55
Draped curtains over solid backing	0.40	0.50	0.60	0.50
Thick carpet on felt underlay	0.30	0.60	0.75	0.80
Expanded polystyrene, 25 mm (1 in.) thick, spaced 50 mm (2 in.) from solid backing	0.55	0.20	0.10	0.15
Acoustic spray plaster, 12 mm ($\frac{1}{2}$ in.) thick, on solid backing	0.50	0.80	0.85	0.60
Metal tiles with 25% perforations, with porous absorbent material laid on top	0.80	0.80	0.90	0.80
Glass wool, 50 mm, on rigid backing	0.50	0.90	0.98	0.99

12.2 Sound absorption in metal foams

Absorption is measured using a plane-wave impedance tube. When a plane sound wave impinges normally on an acoustic absorber, some energy is absorbed and some is reflected. If the pressure p_i in the incident wave is described by

$$p_i = A\cos(2\pi f t) \tag{12.2}$$

and that in the reflected wave (p_r) by

$$p_r = B\cos\left(2\pi f\left(t - \frac{2x}{c}\right)\right) \tag{12.3}$$

then the total sound pressure in the tube (which can be measured with a microphone) is given by the sum of the two. Here f is the frequency (Hz), t is time (s), x is the distance from the sample surface (m), c is the velocity of sound (m/s) and A and B are amplitudes.

The absorption coefficient α is defined as

$$\alpha = 1 - \left(\frac{B}{A}\right)^2 \tag{12.4}$$

It is the fraction of the incident energy of the sound wave which is absorbed by the material. The upper figure shows the value of α for a good absorber, glass wool: at frequencies above 1000 Hz the absorption coefficient is essentially 1, meaning that the sound is almost completely absorbed. The central figure shows absorption in a sample of Alporas foam in the as-received (virgin) state:

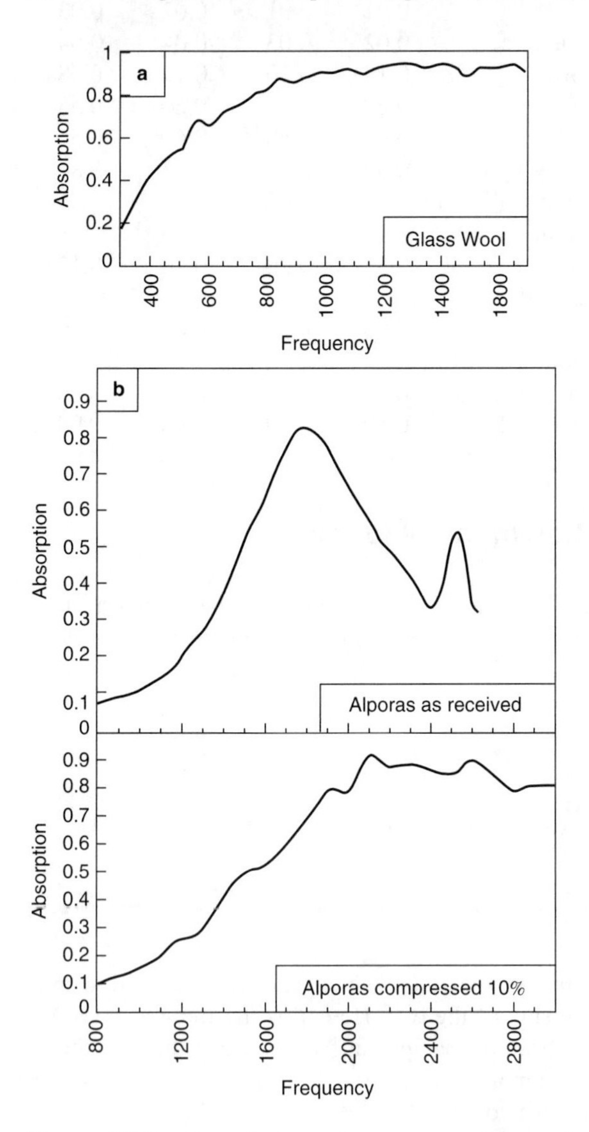

Figure 12.1 *Sound absorption, measured in a plane-wave impedance tube, for glass fiber, Alporas foam in the as-received condition, and Alporas foam after 10% compression to rupture the cell faces*

α rises to about 0.9 at 1800 Hz. Compressing the foam by 10% bursts many of the cell faces, and increases absorption, as shown in the bottom figure. Similar results are reported by Shinko Wire (1996), Asholt (1997), Utsumo *et al.* (1989), Lu *et al.* (1999) and Kovacik *et al.* (1999). In dealing with noise, relative sound level are measured in decibels (dB):

$$\Delta(\mathrm{SPL}) = -10\log_{10}\left(\frac{B}{A}\right)^2 = -10\log_{10}(1-\alpha) \tag{12.5}$$

Thus an absorption coefficient of 0.9 gives a drop in noise level of 10 dB.

The conclusion: metal foams have limited sound-absorbing ability, not as good as glass wool, but still enough to be useful in a multi-functional application.

12.3 Suppression of vibration and resonance

Consider the linear single degree-of-freedom oscillator shown in Figure 12.2(a): a mass m attached by a spring and a damper to a base. Assume that the base vibrates at a single frequency ω with input amplitude X, so that its displacement is $x = Xe^{i\omega t}$. The relative deflection of the mass is $y = Ye^{i\omega t}$ is then given by the transfer function $H(\omega)$:

$$H(\omega) = \frac{Y}{X} = \frac{(\omega/\omega_1)^2}{1-(\omega/\omega_1)^2 + i\eta(\omega/\omega_1)} \tag{12.6}$$

where ω_1 is the undamped natural frequency of the oscillator and η is the damping constant. The magnitude of $H(\omega)$ is shown in Figure 12.2(b).

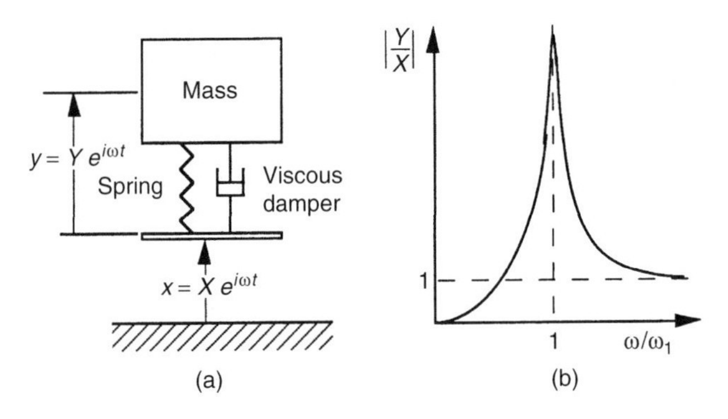

Figure 12.2 *(a) Single degree of freedom oscillator subject to seismic input x at frequency ω. (b) The transfer function for the relative displacement y*

Single low-frequency undamped input

For small values of ω/ω_1 and low damping

$$|Y| = (\omega/\omega_1)^2 \, |X| \tag{12.7}$$

meaning that the response Y is minimized by making its lowest natural frequency ω_1 as large as possible. Real vibrating systems, of course, have many modes of vibration, but the requirement of maximum ω_1 is unaffected by this. Further, the same conclusion holds when the input is an oscillating force applied to the mass, rather than a displacement applied to the base. Thus the material index M_u

$$M_u = \omega_1 \tag{12.8}$$

should be maximized to minimize response to a single low-frequency undamped input.

Consider, as an example, the task of maximizing ω_1 for a circular plate. We suppose that the plate has a radius R and a mass m_1 per unit area, and that these are fixed. Its lowest natural frequency of flexural vibration is

$$\omega_1 = \frac{C_2}{2\pi}\left(\frac{Et^3}{m_1 R^4 (1-\nu^2)}\right)^{1/2} \tag{12.9}$$

where E is Young's modulus, ν is Poisson's ratio and C_2 is a constant (see Section 4.8). If, at constant mass, the plate is converted to a foam, its thickness, t, increases as $(\rho/\rho_s)^{-1}$ and its modulus E decreases as $(\rho/\rho_s)^2$ (Section 4.2) giving the scaling law

$$\frac{\omega_1}{\omega_{1,s}} = \left(\frac{\rho}{\rho_s}\right)^{-1/2} \tag{12.10}$$

– the lower then density, the higher the natural vibration frequency.

Using the foam as the core of a sandwich panel is even more effective because the flexural stiffness, at constant mass, rises even faster as the density of the core is reduced.

Material damping

All materials dissipate some energy during cyclic deformation, through intrinsic material damping and hysteresis. Damping becomes important when a component is subject to input excitation at or near its resonant frequencies.

There are several ways to characterize material damping. Here we use the loss coefficient η which is a dimensionless number, defined in terms of energy dissipation as follows. If a material is loaded elastically to a stress σ_{max} (see

Figure 12.3) it stores elastic strain energy per unit volume, U; in a complete loading cycle it dissipates ΔU, shaded in Figure 12.3, where

$$U = \int_0^{\sigma_{max}} \sigma \, d\varepsilon = \frac{1}{2} \frac{\sigma_{max}^2}{E} \quad \text{and} \quad \Delta U = \oint \sigma \, d\varepsilon \tag{12.11}$$

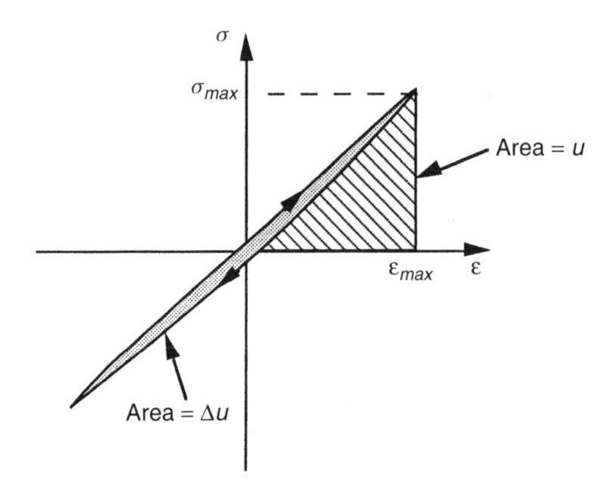

Figure 12.3 *The loss coefficient η measures the fractional energy dissipated in a stress–strain cycle*

The loss coefficient η is the energy loss per radian divided by the maximum elastic strain energy (or the total vibrational energy):

$$\eta = \frac{\Delta U}{2\pi U} \tag{12.12}$$

In general the value of η depends on the frequency of cycling, the temperature and the amplitude of the applied stress or strain.

Other measures of damping include the proportional energy loss per cycle $D = \Delta U/U$, the damping ratio ζ, the logarithmic decrement δ, the loss angle ψ, and the quality factor Q. When damping is small ($\eta < 0.01$) and the system is excited near to resonance, these measures are related by

$$\eta = \frac{D}{2\pi} = 2\zeta = \frac{\delta}{\pi} = \tan\psi = \frac{1}{Q} \tag{12.13}$$

They are no longer equivalent when damping is large.

Broad-band inputs

If the dominant driving frequency is equal to ω_1, then from equation (12.6)

$$|Y| = \frac{1}{\eta}|X| \tag{12.14}$$

and the response is minimized by maximizing the material index:

$$M_d = \eta \qquad (12.15)$$

More generally, the input x is described by a mean square (power) spectral density:

$$S_x(\omega) = S_0 \left(\frac{\omega}{\omega_0}\right)^{-k} \qquad (12.16)$$

where S_0, ω_0 and k are constants, and k typically has a value greater than 2. It can be shown (Cebon and Ashby, 1994) that the material index to be maximized in order to minimise the response to x is

$$M'_d = \eta\omega_1^{k-1} = \eta M_u^{k-1} \qquad (12.17)$$

The selection to maximize M'_d can be performed by plotting a materials selection chart with $\log(\eta)$ on the x-axis and $\log(M_u)$ on the y-axis, as shown in Figure 12.4. The selection lines have slope $1/(1-k)$. The materials which lie farthest above a selection line are the best choice.

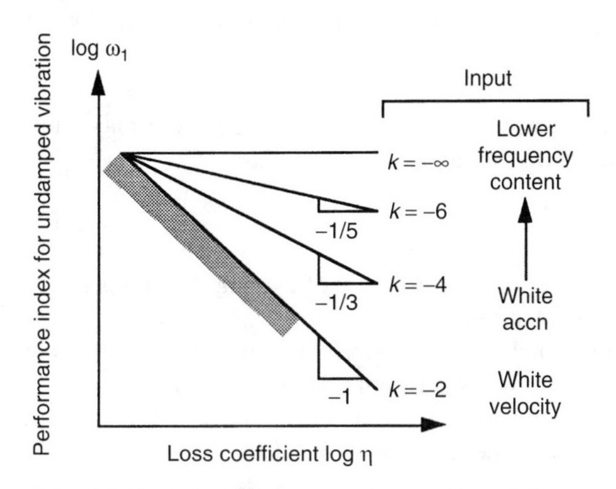

Figure 12.4 *Schematic diagram of a materials selection chart for minimizing the RMS displacement of a component subject to an input with spectral density* $S_0(\omega/\omega_0)^{-k}$

If $k = 0$, the spectrum of input displacement is flat, corresponding to a 'white noise' input, but this is unrealistic because it implies infinite power input to the system. If $k = 2$, the spectrum of the input velocity is flat (or white), which just gives finite power; for this case the selection line on Figure 12.4 has a slope of -1. For larger values of k, the input becomes more concentrated at low frequencies, and the selection line is less steep. If $k = \infty$, the selection line becomes horizontal and the selection task becomes one of choosing materials with the highest value of ω_1, exactly as for the undamped case.

Figure 12.5 shows data for metal foams. Metal foam panels and sandwich panels with metfoam cores have attractive values of M'_d, because of their high flexural stiffness and their relatively high damping capacity. The Alporas range of foams offers particularly good performance. For comparison, aluminum alloys have a values of M_u in the range 1.6–1.8 and damping coefficients, η, in the range $10^{-4} - 2 \times 10^{-3}$ in the same units as those of the figure.

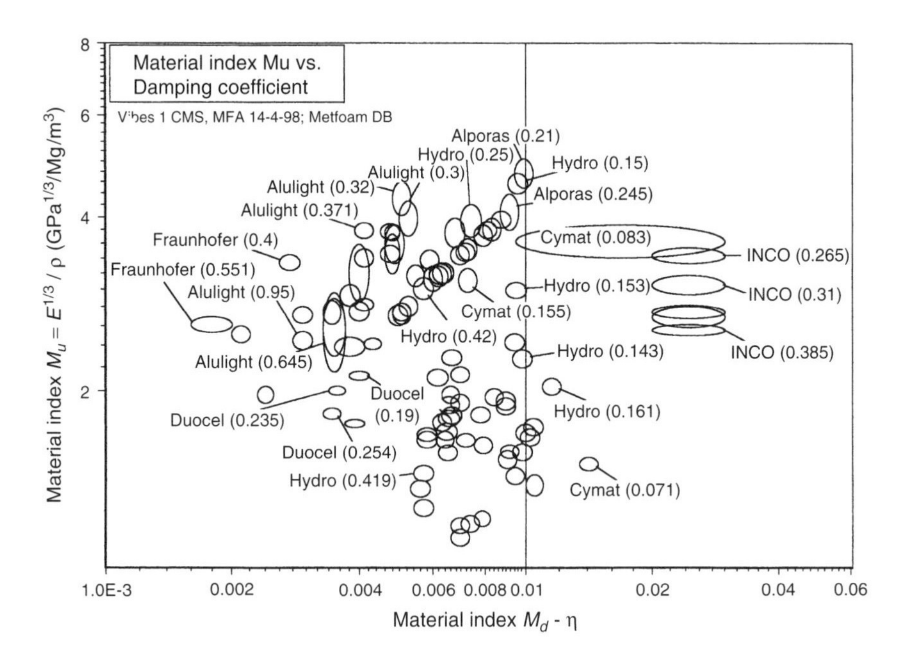

Figure 12.5 *A selection chart for vibration management. The axes are the index. $M_u = E^{1/3}/\rho$ and the damping coefficient $M_d = \eta$. All the materials shown are metal foams, and all have better performance, measured by the index M'_d than the solid metals from which they are made*

References

Asholt, P. (1997) In Banhart, J. (ed.), *Metallschäume*, MIT Verlag, Bremen, Germany.

Beranek, L.L. (ed.) (1960) *Noise Reduction*, McGraw-Hill, New York.

Cebon, D. and Ashby, M.F. (1994) Material selection for precision instruments. *Measurement Science and Technology* **5**, 296–306.

Cowan, H.J. and Smith, P.R. (1988) *The Science and Technology of Building Materials*, Van Nostrand Reinhold, New York.

Kovacik, J., Tobolka, P. Simancik, F. (1999) In Banhart, J., Ashby, M.F. and Fleck, N.A. (eds), *Metal Foams and Foam Metal Structures*, Proc. Int. Conf. *Metfoam'99*, 14–16 June 1999, MIT Verlag, Bremen, Germany.

Lu, T.J., Hess, A. and Ashby, M.F. (1999) *J. Appl. Phys.* **85**, 7528–7539.

Shinko Wire (1996) Alporas Data Sheets (see Section 17: Suppliers of Metal Foams).

Utsuno, H., Tanaka, T. and Fujikawa, T. (1989) *J. Acoust. Soc. Am.* **86**, 637–643.

Wang, X. and Lu, T.J. (1999) Optimised acoustic properties of cellular solids. *J. Accoust. Soc. Am.* **106**(2), 1, 1–10.

Zop, A. and Kollmann, F.G. (1999) In Banhart, J., Ashby, M.F. and Fleck, N.A. (eds), *Metal Foams and Foam Metal Structures*, Proc. Int. Conf. *Metfoam'99*, 14–16 June 1999, MIT Verlag, Bremen, Germany.

Chapter 13

Thermal management and heat transfer

13.1 Introduction

The thermal conductivities of metal foams (see Table 4.1(b)) are at least an order of magnitude greater than their non-metallic counterparts, so they are generally unsuited for simple thermal insulation though they can provide some fire protection. The thermal conductivities of closed-cell foams are, however, lower than those of the fully dense parent metal by a factor of between 8 and 30, offering a degree of fire protection in, for instance, an automobile bulkhead between engine and passenger compartment. More important, open-cell metal foams can be used to enhance heat transfer in applications such as heat exchangers for airborne equipment, compact heat sinks for power electronics, heat shields, air-cooled condenser towers and regenerators (Antohe *et al.*, 1996; Kaviany, 1985). The heat-transfer characteristics of open-cell metal foams are summarized in this chapter.

Examples of the use of metfoams for thermal management can be found in the case studies of Sections 16.4, 16.5, 16.6 and 16.8.

Figure 13.1 illustrates a prototypical heat-transfer configuration. Heat sources are attached to thin conducting substrates between which is bonded a layer of open-celled foam of thickness b and length L. A fluid is pumped at velocity v_f through the foam, entering at temperature T_0 and exiting at temperature T_e. An idealization of the foam structure is shown below: the relative density is $\tilde{\rho} = \rho_c/\rho_s$ and the diameter of the cell edges is d. The local heat-transfer coefficient at the surface of a cell edge is h.

There are three guiding design principles.

1. High conductivity ligaments are needed that transport the heat rapidly into the medium: the preference is for metals such as Cu or Al.
2. A turbulent fluid flow is preferred that facilitates high local heat transfer from the solid surface into the fluid.
3. A low-pressure drop is needed between the fluid inlet and outlet such that the fluid can be forced through the medium using a pumping system with moderate power requirements making low fluid viscosity desirable.

Heat transfer to the fluid increases as either the ligament diameter, d, becomes smaller or the relative density, ρ/ρ_s, increases because the internal

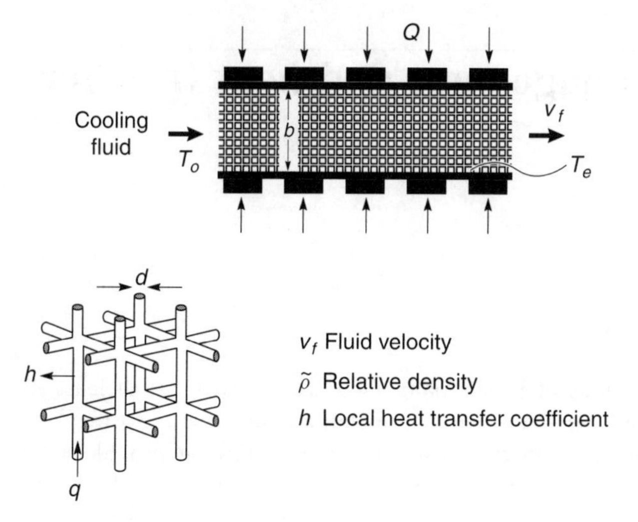

Figure 13.1 *An open-cell foam sandwiched between two conducting plates. Fluid flow from left to right transfers heat from the foam, which has a high surface area per unit volume*

surface area depends inversely on d and the heat conduction cross-section increases with ρ/ρ_s. Counteracting this is the increase in the pressure drop needed to force the fluid through the foam as the surface-area-to-volume ratio increases. Accordingly, for any application there is an optimum cellular structure that depends explicitly on the product specification. These issues are explored more fully below.

13.2 Heat transfer coefficient

The cellular metal is envisaged as a system that transfers heat from a hot surface into a fluid. Thermal performance is characterized by an effective heat transfer coefficient, H_c, which is related to the heat flux, per unit area, q, from the hot surface in the standard manner (e.g. Holman, 1989):

$$q = H_c \Delta T \tag{13.1}$$

where ΔT is a representative temperature drop, roughly equal to the temperature difference between the hot surface and the incoming fluid. A more precise definition is given later. The goal is to develop a cellular system with large H_c that also has acceptable pressure drops and occupies a small volume (compact).

The determination of the heat transfer coefficient, H_c, can be approached in several self-consistent ways. The one presented in this chapter regards the cellular metal as a geometric variant on a staggered bank of cylinders (BOC) oriented normal to the fluid flow. The modified BOC solution has proportionality coefficients that reflect the geometric differences between the foam and the cylinder. This approach has been validated experimentally and the unknown coefficients calibrated. Only the results are given here.

The heat transfer coefficient, H_c, for the cellular metal is (Lu *et al.*, 1998):

$$H_c = \frac{2\tilde{\rho}}{d} k_{eff} \sqrt{Bi_{eff}} \tanh\left[\frac{2b}{d}\sqrt{Bi_{eff}}\right] \tag{13.2}$$

Here $\tilde{\rho} \equiv \rho/\rho_s$, k_{eff} is an effective thermal conductivity related to the actual thermal conductivity of the constituent metal, k_s, by:

$$k_{eff} = 0.28\, k_s \tag{13.3}$$

and b is the thickness of the medium (Figure 13.1). The coefficient of 0.28 has been determined by experimental calibration, using infrared imaging of the cellular medium (Bastawros and Evans, 1997; Bastawros *et al.*, in press). The heat transfer that occurs from the metal ligaments into the fluid can be expressed through a non-dimensional quantity referred to as the Biot number:

$$Bi = \frac{h}{dk_s} \tag{13.4}$$

where h is the local heat transfer coefficient. The Biot number is governed by the dynamics of fluid flow in the cellular medium. The established solutions for a staggered bank of cylinders (e.g. Holman, 1989) are:

$$\begin{aligned} Bi &= 0.91 Pr^{0.36} Re^{0.4} (k_a/k_s) \quad (Re \leqslant 40) \\ &= 0.62 Pr^{0.36} Re^{0.5} (k_a/k_s) \quad (Re > 40) \end{aligned} \tag{13.5}$$

where Re, the Reynolds number, is

$$Re = \frac{v_f d}{\nu_a} \tag{13.6}$$

with v_f the free stream velocity of the fluid, ν_a its kinematic viscosity, k_a its thermal conductivity and Pr is the Prandtl number (of order unity). For the cellular metal, Bi will differ from equation (13.5) by a proportionality coefficient (analogous to that for the thermal conductivity) resulting in an effective value:

$$Bi_{eff} = 1.2\, Bi \tag{13.7}$$

where the coefficient 1.2 has been determined by experimental calibration (Bastawros *et al.*, in press).

This set of equations provides a complete characterization of the heat transfer coefficient. The trends are found upon introducing the properties of the foam (d, ρ/ρ_s and k_s), its thickness b, and the fluid properties (ν_a, k_a and Pr), as well as its velocity v_f. The caveat is that the proportionality constants in equations (13.3) and (13.7) have been calibrated for only one category of open cell foam: the DUOCEL range of materials. Open-cell foams having different morphology are expected to have different coefficients. Moreover, if ρ/ρ_s and d are vastly different from the values used in the calibration, new domains of fluid dynamics may arise, resulting again in deviations from the predictions.

The substrate attached to the cellular medium also contributes to the heat transfer. In the absence of a significant thermal constriction, this contribution may be added to H_c. Additional interface effects can reduce H_c, but we shall ignore these.

13.3 Heat fluxes

The heat, Q, flowing into the fluid through the cellular medium per unit width is related to the heat transfer coefficient by:

$$Q = LH_c \Delta T_{\ell m} \tag{13.8}$$

where L is the length of the foam layer in Figure 13.1. Here $\Delta T_{\ell m}$ is the logarithmic mean temperature. It is related to the temperature of the heat source T_1 as well as fluid temperature at the inlet, T_0, and that at the outlet, T_e by:

$$\Delta T_{\ell m} = \frac{T_e - T_0}{\ell n[(T_1 - T_0)/(T_1 - T_e)]} \tag{13.9}$$

Usually, T_1 and T_0 are specified by the application. Accordingly, T_e must be assessed in order to determine Q. For preliminary estimates, the approximation

$$\Delta T_{\ell m} \approx T_1 - T_0 \tag{13.10}$$

may be used. Explicit determination requires either experimental measurements or application of the following expressions governing the fluid flows.

The temperature in the fluid along the x-direction varies as

$$T_f = T_1 - (T_1 - T_0)\exp(-x/\ell) \tag{13.11}$$

where ℓ is a transfer length governed by the properties of the cellular metal, the fluid and the substrate. In the absence of a thermal resistance at the attachments, this length is:

$$\ell = \frac{\rho_a c_p b v_f}{2\eta k_{eff}\sqrt{Bi_{eff}}}\left[1 + \frac{\tilde{\rho}}{1.5\eta}\tanh\frac{2b}{d}\sqrt{Bi_{eff}}\right]^{-1} \tag{13.12}$$

where c_p is the specific heat of the fluid and $\eta = 1 - 0.22(\rho/\rho_s)$. The exit temperature may thus be determined by introducing ℓ from equation (13.12) into (13.11) and setting $x = L$, whereupon $T_f \equiv T_e$.

Expected trends in the heat flux, Q, dissipated by cellular metals (in W/m^2) can be anticipated by using the above formulae. Typically, this is done using non-dimensional parameters, as plotted in Figure 13.2 with air as the cooling fluid. The parameters are defined in Table 13.1. The principal feature is the substantial increase in heat dissipation that can be realized upon either decreasing the cell edge diameter, d, or increasing the relative density, ρ/ρ_s. Eventually, a limit is reached, governed by the heat capacity of the cooling fluid.

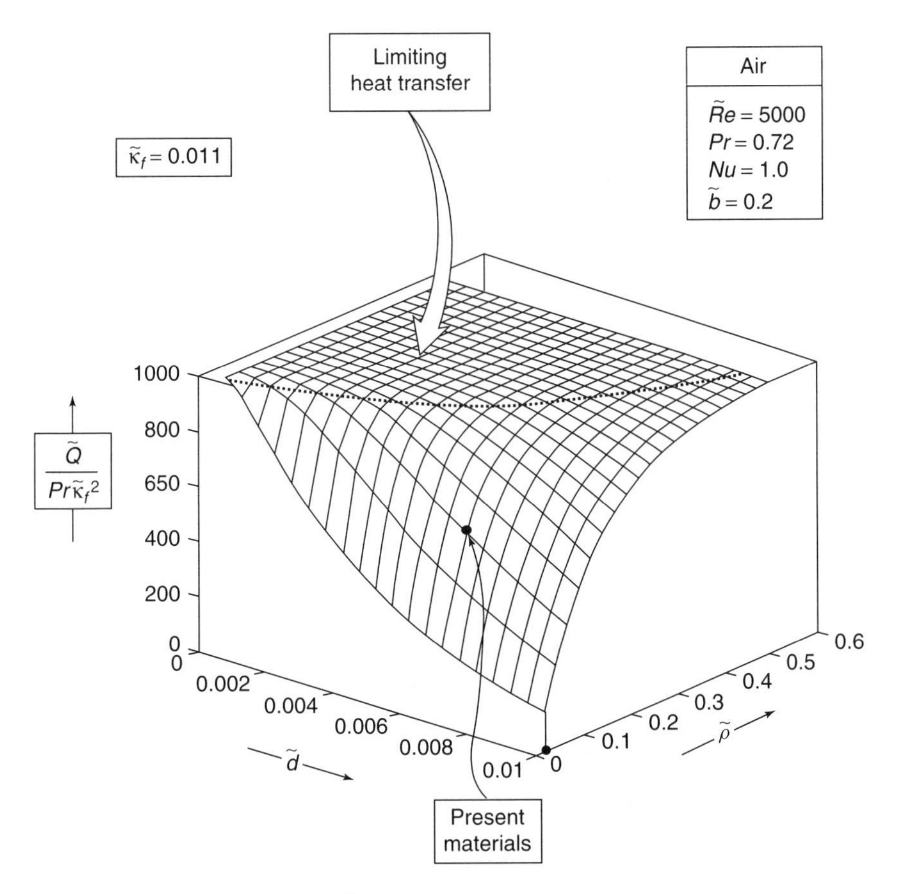

Figure 13.2 *The heat flux $\tilde{Q} = Q/k_s[T_I - T_O]$ into the fluid, plotted as a function of the relative density, $\tilde{\rho}$, and the dimensionless cell-edge diameter, $\tilde{d} = d/L$*

Table 13.1 *Non-dimensional parameters for cellular metal heat dissipation*

Heat flux	$\tilde{Q} = Q/k_s[T_1 - T_0]$
Prandtl number	$\mathrm{Pr} = \nu_a/\alpha_a$
Reynolds number	$\tilde{Re} = v_f L/\nu_a$
Cell wall thickness	$\tilde{d} = d/L$
Foam thickness	$\tilde{b} = b/L$
Nusselt number	$Nu = Bik_s/k_a$
Thermal conduction	$\tilde{K}_f = \sqrt{k_a/k_s}$
Power dissipation	$\tilde{P} = \Delta p v_f b L^2/\rho_a \nu_a^3$

α_a = thermal diffusivity of cooling fluid.

13.4 Pressure drop

As the heat transfer coefficient increases, so does the pressure drop across the medium. The latter can sometimes be the limiting factor in application, because of limitations on the available pumping power. The pressure drop Δp has the general form

$$\Delta p/L = \xi(1/a)[\nu_a^m \rho_a/(1-\alpha)^{2-m}]v_f^{2-m} d^{-m} \tag{13.13}$$

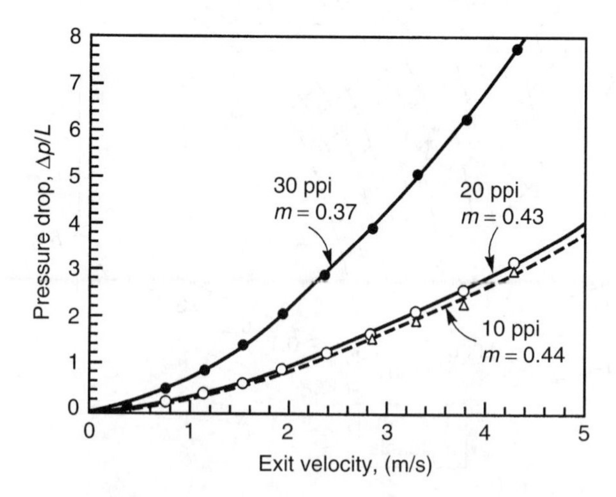

Figure 13.3 *The pressure drop per unit length of flow plotted against the exit velocity for Duocel foams, from which the power dissipated in pumping the fluid can be estimated*

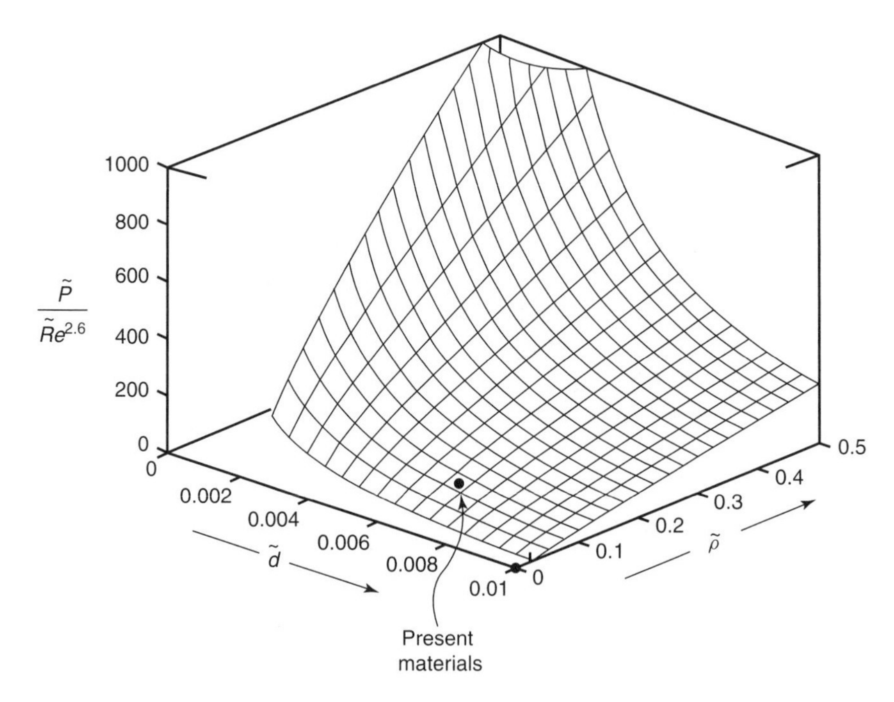

Figure 13.4 *The power dissipation plotted as a function of the relative density, $\tilde{\rho}$, and the dimensionless cell-edge diameter, $\tilde{d} = d/L$*

where a is the cell size

$$a = 1.24d\sqrt{3\pi/\tilde{\rho}} \tag{13.14}$$

The exponent m and the coefficient ξ have been calibrated by experimental measurements. They are:

$$m = 0.4$$
$$\xi = 4 \tag{13.15}$$

Some typical results are plotted in Figure 13.3. Pressure drops for other conditions can be predicted from equations (13.13) to (13.15), again with the proviso that the fluid flow scaling (13.5) retains its validity. The expected behavior is illustrated in Figure 13.4.

13.5 Trade-off between heat transfer and pressure drop

For any system there is a trade-off between heat flux and pressure drop. A cross-plot of these two quantities in accordance with the non-dimensional

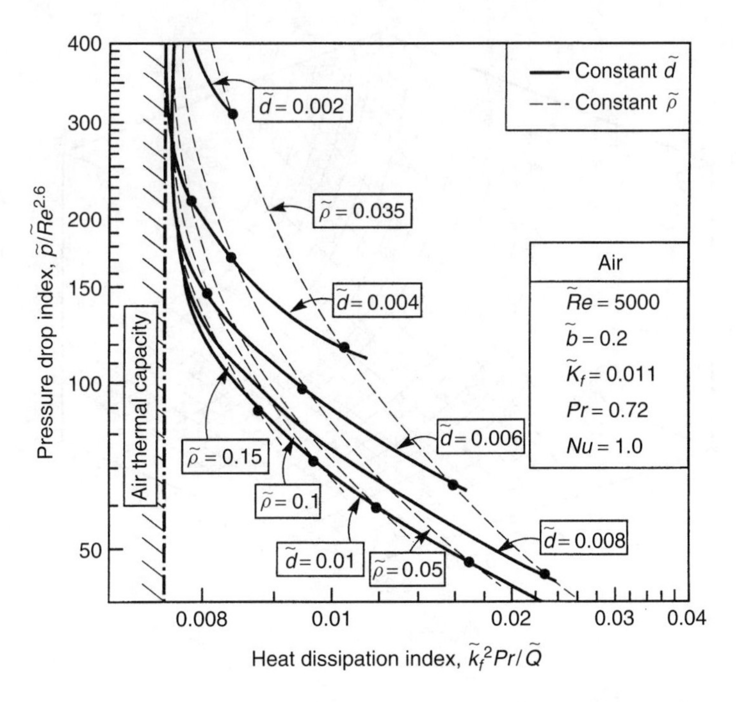

Figure 13.5 *The trade-off between pressure drop index and heat dissipation index*

parameters defined by the model and defined in Table 13.1 illustrates this trade-off. It is shown in Figure 13.5.

References

Antohe, B.V., Lage, J.L., Price, D.C. and Weber, R.M. (1996) *Int. J. Heat Fluid Flow* **17**, 594–603

Bastawros, A.-F., and Evans, A.G. (1997) *Proc. Symp. on the Applications of Heat Transfer in Microelectronics Packaging*, IMECE, Dallas, Texas, USA.

Bastawros, A.-F., Stone, H.A. and Evans, A.G. (in press) *J. Heat Transfer.*

Holman, J.P. (1989) *Heat Transfer – a Modern Approach*, McGraw-Hill, New York.

Kaviany, M. (1985) *Int. J. Heat Mass Transfer* **28**, 851–858.

Lu, T.J., Stone, H.A. and Ashby, M.F. (1998) *Acta Materialia* **46**, 3619–3635.

Chapter 14

Electrical properties of metal foams

The electrical conductivity of a metal foam is less than that of the metal from which it is made for the obvious reason that the cell interiors, if gas-filled, are non-conducting. One might guess that the conductivity should vary linearly with the relative density, but the real dependence is stronger than linear, for reasons explained below. Though reduced, the conductivity of metal foams is more than adequate to provide good electrical grounding and shielding of electromagnetic radiation.

The large, accessible surface area of open cell metal foams makes them attractive as electrodes for batteries (see Case Study 17.6). Nickel foams are extensively used in this application.

14.1 Measuring electrical conductivity or resistivity

The electrical resistivity of a thick metal foam sheet can be measured using a four-point probe technique sketched in Figure 14.1. Two probes (P1 and P4) are used to introduce a current, I, into the sample while a pair of different probes (P2 and P3) are used to measure the potential drop, V, between them. If the plate is sufficiently thick, the electrical resistivity, Θ, of the foam (commonly measured in units of $\mu\Omega.\text{cm}$) is given by:

$$\Theta = 2\pi\left(\frac{V}{IS}\right) \tag{14.1}$$

where

$$S = \frac{1}{s_1} + \frac{1}{s_3} - \frac{1}{s_1 + s_2} - \frac{1}{s_2 + s_3} \tag{14.2}$$

and s_1, s_2 and s_3, are the probe spacings shown in the figure.

The electrical conductivity, σ (units, $\Omega^{-1}.\text{m}^{-1}$), is the reciprocal of the resistivity. The resistance, R, of a piece of foam of length ℓ and a cross-sectional area A normal to the direction of current flow is given by:

$$R = \Theta\frac{\ell}{A} = \frac{\ell}{\sigma A} \tag{14.3}$$

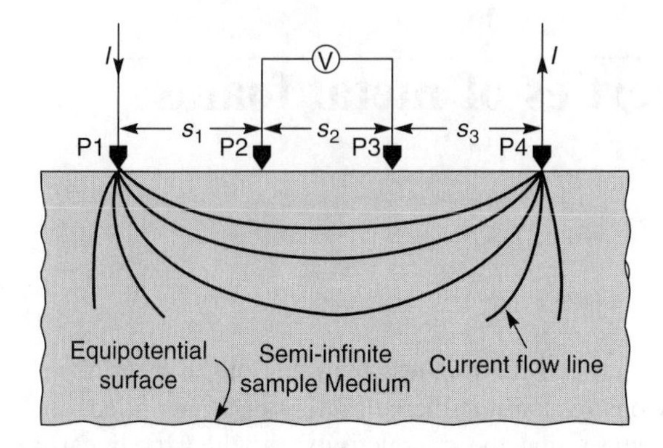

Figure 14.1 *A four-point probe method for measuring the electrical conductivity of metal foams*

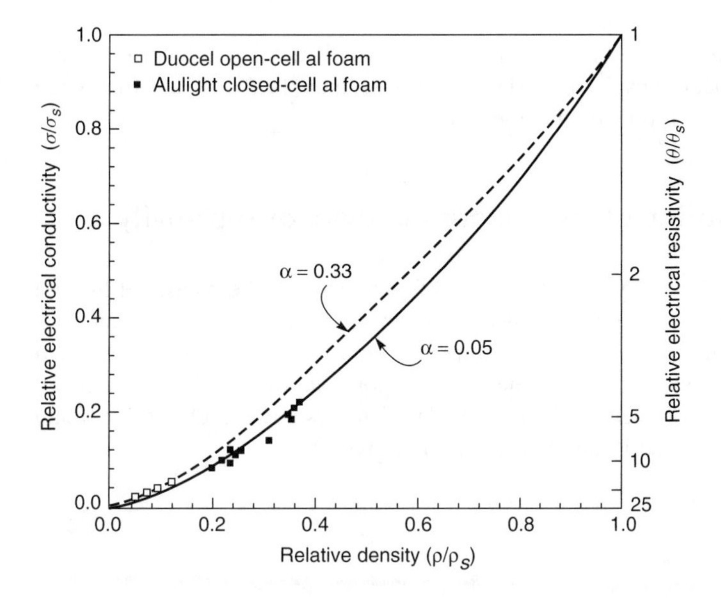

Figure 14.2 *The relative electrical resistivity and conductivity of open- and closed-cell aluminum foams of differing relative density. The lines are plots of equation (14.4) with $\alpha = 0.33$ and $\alpha = 0.05$*

14.2 Data for electrical resistivity of metal foams

Little data for the electrical resistivity of metal foams has been reported. Figure 14.2 shows measurements of the conductivity of open (ERG-Duocel) and closed (Mepura-Alulight) cell aluminum foams as a function of relative density, normalized by the conductivity of the fully dense alloy. The conductivity varies in a non-linear way with relative density. Additional data for nickel foams can be found in patents referenced as Babjak *et al.* (1990), at the end of this chapter.

14.3 Electrical conductivity and relative density

Figure 14.3 shows an idealization of a low-density open cell foam. The cell edges have length ℓ and cross-section $t \times t$, meeting at nodes of volume t^3. The relative density of an open-cell foam is related to the dimensions of the cells (omitting terms of order $(t/\ell)^4$) by

$$\frac{\rho}{\rho_s} \approx \frac{3t^2}{\ell^2} \tag{14.4}$$

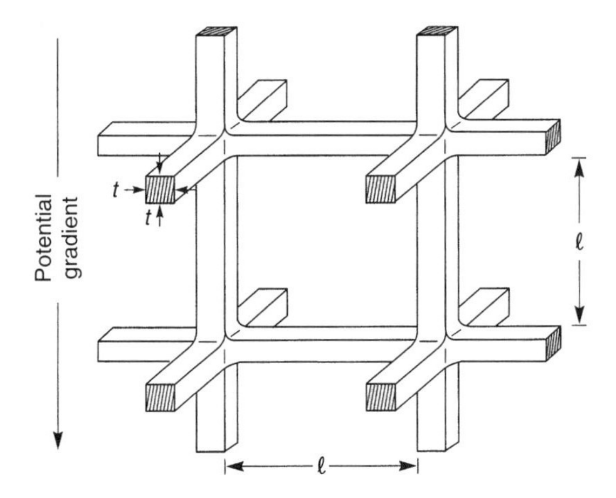

Figure 14.3 *An idealized open-cell foam consisting of cell edges of length ℓ and cross-section t^2, meeting at nodes of volume t^3. In real foams the nodes are larger ('Plateau borders') and the edges thinner at their mid-points, because of the effects of surface tension*

The dependence of electrical conductivity on relative density can be understood in the following way. The cell edges form a three-dimensional network. If a potential gradient is applied parallel to one set of cell edges, the edges which lie parallel to the gradient contribute to conduction but those which lie normal to it do not, because there is no potential difference between their ends. The network is linked at nodes, and the nodes belong to the conducting path. At low relative densities the volume of the nodes is negligible compared with that of the edges, and since only one third of these conduct in a body containing a fraction ρ/ρ_s of conducting material, the relative conductivity is simply

$$\frac{\sigma}{\sigma_s} = \frac{1}{3}\left(\frac{\rho}{\rho_s}\right) \tag{14.5}$$

where σ_s is the conductivity of the solid from which the foam was made.

As the relative density increases, the nodes make an increasingly large contribution to the total volume of solid. If the node volume scales as t^3 and that of the edges as $t^2\ell$, then the relative contribution of the nodes scales as t/ℓ, or as $(\rho/\rho_s)^{1/2}$. We therefor expect that the relative conductivity should scale such that

$$\begin{aligned}\frac{\sigma}{\sigma_s} &= \frac{1}{3}\left(\frac{\rho}{\rho_s}\right)\left(1 + 2\left(\frac{\rho}{\rho_s}\right)^{1/2}\right) \\ &= \frac{1}{3}\left(\frac{\rho}{\rho_s}\right) + \frac{2}{3}\left(\frac{\rho}{\rho_s}\right)^{3/2}\end{aligned} \tag{14.6}$$

where the constant of proportionality 2 multiplying $(\rho/\rho_s)^{1/2}$ has been chosen to make $\sigma/\sigma_s = 1$ when $\rho/\rho_s = 1$, as it obviously must.

Real foams differ from the ideal of Figure 14.3 in many ways, of which the most important for conductivity is the distribution of solid between cell edges and nodes. Surface tension pulls material into the nodes during foaming, forming thicker 'Plateau borders', and thinning the cell edges. The dimensionality of the problem remains the same, meaning that the fraction of material in the edges still scales as ρ/ρ_s and that in the nodes as $(\rho/\rho_s)^{3/2}$, but the multiplying constants depend on precisely how the material is distributed between edges and nodes. We therefore generalize equation (14.6) to read

$$\frac{\sigma}{\sigma_s} = \alpha\left(\frac{\rho}{\rho_s}\right) + (1-\alpha)\left(\frac{\rho}{\rho_s}\right)^{3/2} \tag{14.7}$$

retaining the necessary feature that $\sigma/\sigma_s = 1$ when $\rho/\rho_s = 1$. This means that the conductivity of a foam can be modeled if one data point is known, since this is enough to determine α.

Equation (14.7) is plotted on Figure 14.2, for two values of α. The upper line corresponds to $\alpha = 0.33$ (the 'ideal' behavior of equation (14.6)), and describes the open-cell Duocell results well. The lower line corresponds to $\alpha = 0.05$, meaning that the edges make a less than ideal contribution to conductivity. It fits the data for Alulight well.

Alulight is an closed-cell foam, yet its behavior is describe by a model developed for open cells. This is a common finding: the moduli and strengths of closed-cell foams also lie close to the predictions of models for those with open cells, perhaps because the cell faces are so thin and fragile that they contribute little to the properties, leaving the edges and nodes to determine the response.

References

Babjak, J., Ettel, V.A., Passerin, V. (1990) US Patent 4,957,543.

Gibson, L.J. and Ashby, M.F. (1997) *Cellular Solids, Structure and Properties*, 2nd edition, p. 296, Cambridge University Press, Cambridge.

Kraynik, A.M., Neilsen, M.K., Reinelt, D.A. and Warren, W.E. (1997) In Sadoc, J.F. and River, N. (eds), *Foam Micromechanics, Foams and Emulsion*, p. 259, Proceedings NATO Advanced Study Institute, Cargese, Corsica, May.

Lemlich, R. (1997) *J. Colloid and Interface Science* **64**(1), 1078.

Mepura (1995) Data Sheets, Metallpulvergesellschaft m.b.h. Randshofen A-5282, Braunau, Austria.

Phelan, R., Weaire, D., Peters, E. and Verbist, G. (1996) *J. Phys: Condens Matter* **8**, L475.

Chapter 15

Cutting, finishing and joining

The cellular structure of metal foams requires that special techniques are required to give high-quality cuts and joints:

- Rough cutting is practical with conventional machine tools, but with some surface damage.
- High-quality surfaces require the use of electro-discharge machining, chemical milling, water-jet cutting or high-speed machining.
- Adhesives work well as bonding agents for metal foams.
- Welding, brazing and soldering are all possible.
- Threaded, embedded and bolted fasteners require careful design if they are to function well, and are sensitive to fatigue loading.

Guidelines for the design and use of joining methods are assembled in this chapter.

15.1 Cutting of metal foams

Conventional cutting and machining techniques (bandsawing, milling, drilling) cause severe surface distortion or damage to low-density metal foams. Accurate cutting is possible by electro-discharge machining (EDM), by chemical milling, by water-jet cutting or by the use of very high-speed fly-cutters. When making test samples of metal foams for characterization (see Chapter 3) it is important to use electro-discharge machining or chemical milling unless the samples are very large, because surface damage caused by other cutting methods influences the properties.

15.2 Finishing of metal foams

The cut surface of a metal foam has open cells, a rough texture, and is vulnerable to local damage. The surface can be filled with an epoxy or other resin, or clad (creating a sandwich structure) with a skin of a material compatible with that of the foam. Syntactic structures have a natural skin, which can be polished, etched, anodized or coated by conventional methods.

15.3 Joining of metal foams

Metal foams can be soldered and welded. Foams have a cellular structure resembling, in some ways, that of wood. Because of this they can be joined in ways developed for wood, using wood screws, glue joints or embedded fasteners. Figure 15.1 summarizes joining methods.

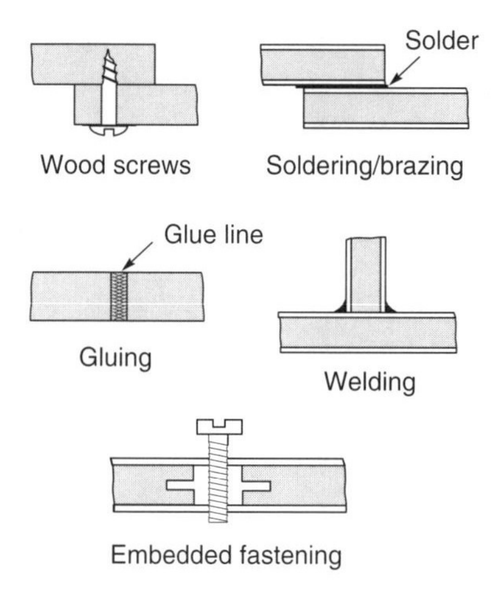

Figure 15.1 *Ways of fastening and joining metal foams*

Welding, brazing and soldering

Welding and brazing are best used for foams with integral skins. Studies of laser welding (Burzer *et al.*, 1998) show promise, but the technique requires careful control. Brazing of Al-based foams with aluminum–eutectic alloys is practical. The soldering of aluminum foams requires a flux to remove the oxide film. If the flux penetrates the foam it causes corrosion, so soldering is only practical for sandwiches or skinned structures, restricting the solder to the outer surface of the skin. Soldered joints weaken the foam, which fails at a stress less that the tensile strength of the foam itself.

Adhesives

Foams can be glued with the same adhesives used to bond the base metal (Olurin, *et al.*, 1999). The glue joints are usually stronger than the foam itself. There are some drawbacks: low thermal stability, mismatch of expansion coefficient and the possible creation of a thermal and electrical isolation barrier. Provided these are not critical to the design, adhesives (particularly epoxies) allow simple, effective attachment. Typical of their use is the attachment of face-sheets to metal foam cores in sandwich-panel construction.

Fasteners

Embedded fasteners (Figure 15.2), when strongly bonded by threads or adhesives to the foam itself, pull out at an axial load

$$F_f = 2\pi R(\ell - x)\sigma_s \tag{15.1}$$

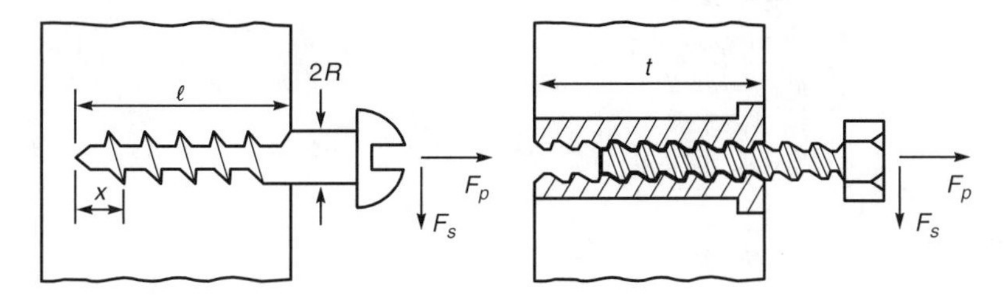

Figure 15.2 *Embedded fasteners: wood screws and inserts*

where $2R$ is the diameter, ℓ is the embedded length of the fastener, $x \approx R$ is a small end correction allowing for the tapered tip of the fastener if there is one, and σ_s is the shear-yield strength of the foam. This last can be estimated as

$$\sigma_s = 0.2\left(\frac{\rho}{\rho_s}\right)^{3/2}\sigma_{y,s} \tag{15.2}$$

giving

$$F_f \approx 0.4\pi R(\ell - x)\left(\frac{\rho}{\rho_s}\right)^{3/2}\sigma_{y,s} \tag{15.3}$$

where $\sigma_{y,s}$ is the yield strength of the solid of which the foam is made. Figure 15.3 shows the dependence of pull-out load on foam density, on logarithmic scales, for a range of fasteners. All scale with density in the way described by Equation (15.3). Experiments confirm the dependence on length and diameter.

Bolted fasteners (Figure 15.4) fail when the head of the fastener pulls through the foam. The pull-through load is

$$F_p = \pi(R_w^2 - R^2)\sigma_c + 2\pi R_w \gamma \tag{15.4}$$

The first term is simply the crushing strength, σ_c, of the foam times the contact area of the washer, of radius R_w. The second accounts for the tearing of the

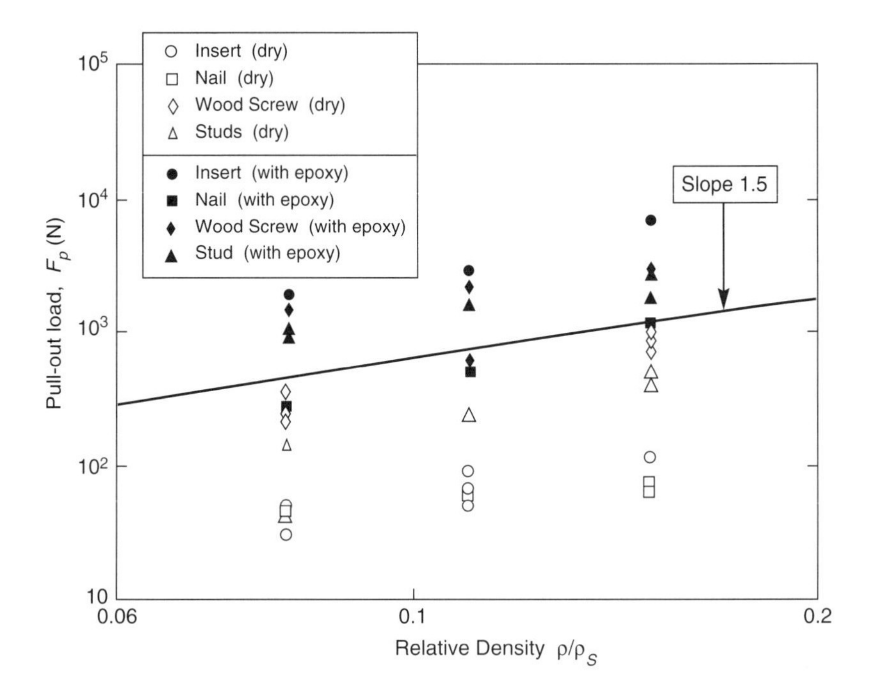

Figure 15.3 *The dependence of pull-out load on relative density for a range of embedded fastners (wood screw diameter $2a = 4.8$ mm; nail diameter $2a = 4.5$ mm; stud diameter $2a = 6$ mm; threaded insert diameter $2a = 20$ mm; all have embedded length $\ell = 20$ mm)*

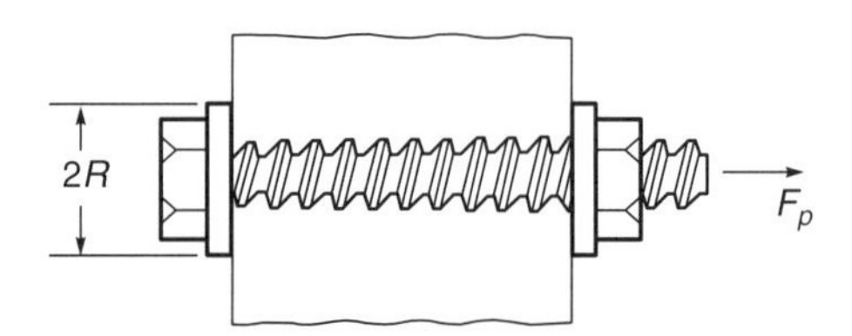

Figure 15.4 *A bolted fastener*

foam around the periphery of the washer. The tear-energy per unit area, γ, has been measured. It is adequately described by

$$\gamma \approx \left(\frac{\rho}{\rho_s}\right)^{3/2} \gamma_0 \qquad (15.5)$$

where γ_0 is a characteristic of the material of which the foam is made; for Alporas foam, its value is $\gamma_0 = 260\,\text{kJ/m}^2$.

Through-fasteners may be required to carry bearing loads (Figure 15.5). Initial yield occurs when the mean bearing load exceeds the crushing strength of the foam, that is, when

$$F_b = 2Rt\sigma_c \tag{15.6}$$

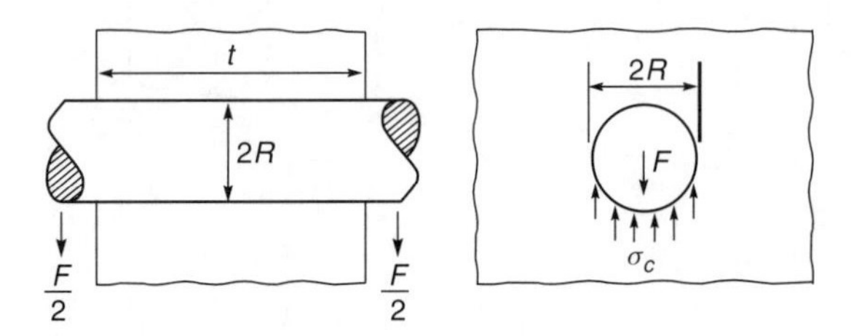

Figure 15.5 *A through-fastener carrying bearing loads*

where σ_c is the crushing strength of the foam, approximated by

$$\sigma_c \approx 0.3 \left(\frac{\rho}{\rho_s}\right)^{3/2} \sigma_{y,s}$$

(see Chapter 4, Equation (4.2)). Once yielding has occurred the fastener is no longer secure and fretting or local plasticity caused by tilting make its behavior unpredictable. Equation (15.6), with an appropriate safety factor, becomes the safe design criterion.

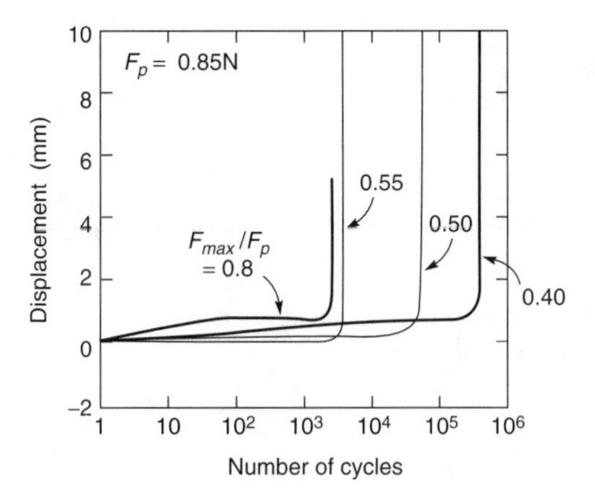

Figure 15.6 *The accumulated displacement under tension–tension cyclic loading of embedded fastners at various levels of peak cyclic stress*

Cyclic loading of fasteners

Cyclic loading leads to a response typified by Figure 15.6. At peak cyclic loads, F_{max}, below the monotonic pull-out load, F_f, the response (both for pull-out, Equation (15.3), and for bearing loads, Equation (15.6)) is essentially elastic until, at a critical number of cycles which depends on F_{max}/F_f, the displacement per cycle increases dramatically. This response parallels that for fatigue of plain specimens (Chapter 8). Figure 15.7 shows a plot of load, normalized by the monotonic pull-out load, versus number of cycles to failure, defined as the number of cycles at the knee of the curves of Figure 15.6, for an embedded fastner subjected to cyclic pull-out loads. It resembles the S–N curves for metal foams, but the slope is slightly steeper, suggesting that damage accumulates slightly faster than in the cyclic loading of plain specimens.

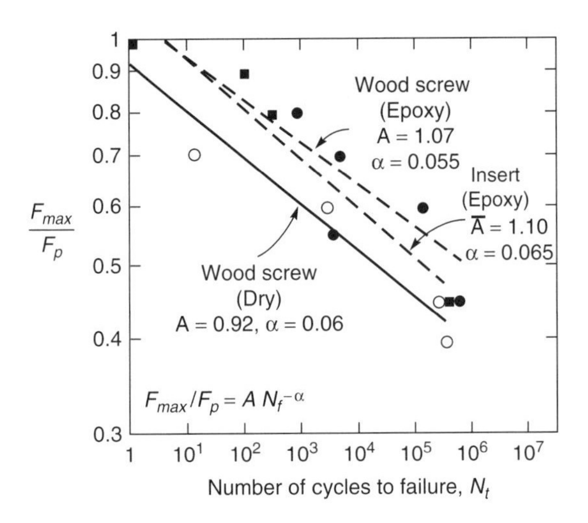

Figure 15.7 *Cyclic peak-load plotted against number of cycles to failure for embedded fasteners*

References

Burzer, J., Bernard, T. and Bergmann, H.W. (1998) Joining of aluminum structures with aluminum foams. In Shwartz, D.S., Shih, D.S., Evans, A.G. and Wadley, H.N.G. (eds), *Porous and Cellular Materials for Structural Application*, Materials Research Society Proceedings, Vol. 521, MRS, Warrendale, PA, USA

Olurin, O.B. Fleck, N.A. and Ashby, M.F. (1999) Joining of aluminum foams with fasteners and adhesives. To appear in *J. Mat. Sci.*

Sedliakova, N., Simancik, F., Kovacik, J. and Minar, P (1997) In Banhart, J. (ed), *Metallschäume*, Fraunhofer Institut für Angewandte Materialforschung, Bremen, MIT Press, pp. 177–185.

Chapter 16

Cost estimation and viability

When are metal foams viable? By viable we mean that the balance between performance and cost is favourable. The answer has three ingredients: a technical model of the performance of the material in a given application, a cost model giving an estimate of material and process costs, and a value model which balances performance against cost. Viability is assessed by constructing a value function that includes measures of both performance and cost. It allows ranking of materials by both economic and technical criteria.

At present all metal foams are produced in small quantities using time and labor-intensive methods, and all, relative to the solid metals from which they derive, are expensive. But it is not the present-day cost which is relevant; it is the cost which would obtain were the process to be scaled and automated to meet the increases demand of one or a portfolio of new applications. The role of a cost model is to assess this, to identify cost drivers, to examine the ultimate limits to cost reduction, and to guide process development.

Balancing cost against performance is an example of multi-objective optimization. This chapter expands on this, and the role of cost and performance metrics in determining viability. The method, which can encompass multifunctionality, is illustrated by examples.

16.1 Introduction: viability

The viability of a foam in a given application depends on the balance between its performance and its cost. There are three steps in evaluating it (Figure 16.1).

The first is the technical assessment (Figure 16.1, upper circle). Performance metrics are identified and evaluated for competing solutions for the design. Examples of technical models can be found in Chapter 5.

The second step is the analysis of cost (Figure 16.1, lower-left circle): how much does it cost to achieve a given performance metric? The quantity of foam required to meet constraints on stiffness, on strength, on energy absorption, etc. is calculated from straightforward technical models. The cost, C, of producing this quantity of material in the desired shape is the output of the cost model, as described in Section 16.3.

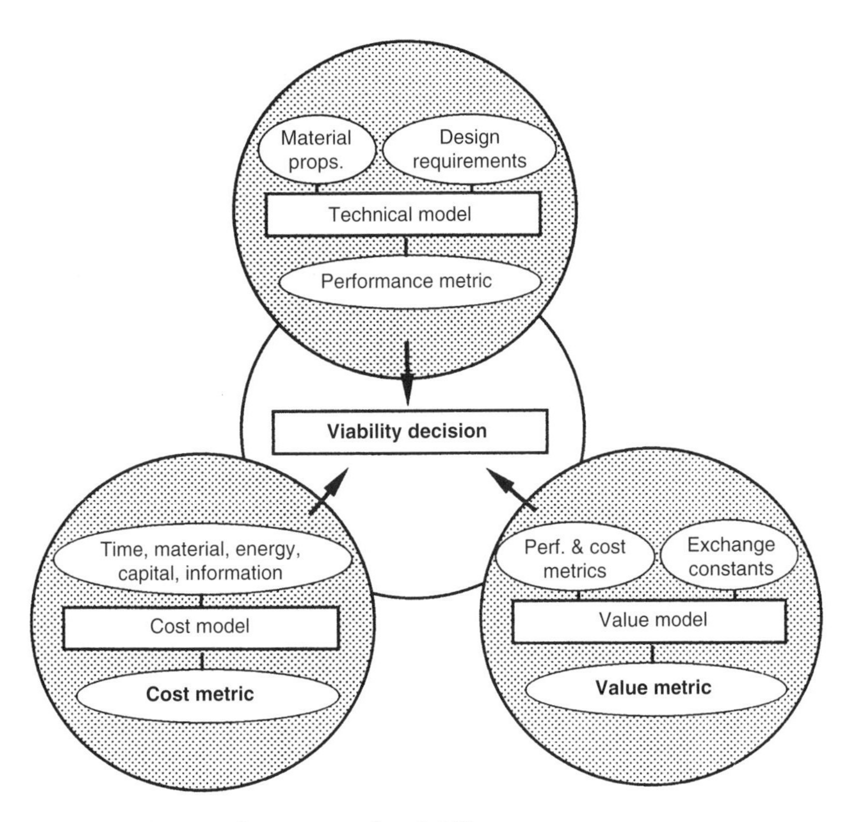

Figure 16.1 *The three parts of a viability assessment*

The final step is that of assessing value (Figure 16.1, lower-right circle): is the change in performance worth the change in cost? Balancing performance against cost is an example of multi-objective optimization. This is discussed in Section 16.4

16.2 Technical modeling and performance metrics

To construct a technical model, performance metrics P_i, are identified and evaluated for the foam and for competing materials or systems. A performance metric is a measure of the performance offered by the material in a particular application. In minimum weight design the performance metric is the mass: the lightest material which meets the specifications on stiffness, strength, etc. is the one with the greatest performance. In design for energy-mitigation, in which it is desired that a protective packaging should crush, absorbing a specified energy, and occupy as little volume as possible, the performance metric is the volume. In design to minimize heat loss, the metric might be the heat flux per unit area of structure. In design for the environment, the metric is a measure of the environmental load associated with the manufacture, use and disposal of the material. Each performance metric defines an objective: an attribute of the structure that it is desired to maximize or minimize. Models are developed for the performance metrics following the method of Chapter 5; the resulting performance equations contain groups of material properties or 'material indices'. The material that maximizes the chosen performance metric has the highest technical merit. But for this to be a viable choice, the performance must be balanced against the cost.

16.3 Cost modeling

The manufacture of a foam (or of a component made from one) consumes resources (Table 16.1). The final cost is the sum of these resources. This resource-based approach to cost analysis is helpful in selecting materials and processes even when they differ greatly, since all, no matter how different, consume the resources listed in the table. Thus the cost of producing one kg of foam entails the cost C_m ($/kg) and mass, m, of the materials and feedstocks from which it is made, and it involves the cost of dedicated tooling, C_t ($), which must be amortized by the total production volume, n (kg). In addition, it requires time, chargeable at an overhead rate $\dot{C}_L$ (with units of $/h or equivalent), power $\dot{P}$ (kW) at an energy cost C_e ($/kW.h), space of area

Table 16.1 *The resources consumed in making a material*

Resource		*Symbol*	*Unit*
Materials:	inc. consumables	C_m	$/kg
Capital:	cost of equipment	C_c	$/unit
	cost of tooling	C_t	$/unit
Time:	overhead rate	$\dot{C}_L$	$/hr
Energy:	power	$\dot{P}$	kW
	cost of energy	C_e	$/kW.h
Space:	area	A	m^2
	cost of space	$\dot{C}_S$	$/m^2.h
Information:	R&D	C_i	
	royalty payments		

A, incurring a rental cost of $\dot{C}_S$ (\$/m^2.h), and information, as research and development costs, or as royalty or licence payments $\dot{C}_i$ (expressed as \$/h). The cost equation, containing terms for each of these, takes the form

$$\begin{array}{cccccc} \text{[Material]} & \text{[Tooling]} & \text{[Time]} & \text{[Energy]} & \text{[Space]} & \text{[Information]} \\ \downarrow & \downarrow & \downarrow & \downarrow & \downarrow & \downarrow \end{array}$$

$$C = \left[\frac{mC_m}{1-f}\right] + \left[\frac{C_t}{n}\right] + \left[\frac{\dot{C}_L}{\dot{n}}\right] + \left[\frac{\dot{P}C_e}{\dot{n}}\right] + \left[\frac{A\dot{C}_S}{\dot{n}}\right] + \left[\frac{\dot{C}_i}{\dot{n}}\right] \tag{16.1}$$

where $\dot{n}$ is the production rate (kg/h) and f is the scrap rate (the material wastage in the process).

A given piece of equipment – a powder press, for example – is commonly used to make more than one product, that is, it is not dedicated to one product alone. It is usual to convert the capital cost, C_c, of non-dedicated equipment, and the cost of borrowing the capital itself, into an overhead by dividing it by a capital write-off time, t_c (5 years, say) over which it is to be recovered. Thus the capital-inclusive overhead rate becomes

$$\begin{array}{cc} \text{[Basic OH rate]} & \text{[Capital write-off]} \\ \downarrow & \downarrow \end{array}$$

$$\frac{\dot{C}_L}{\dot{n}} = \frac{1}{\dot{n}}\left\{\left[\dot{C}_{L0}\right] + \left[\frac{C_c}{Lt_c}\right]\right\} \tag{16.2}$$

where $\dot{C}_{L0}$ is the basic overhead rate (labor, etc.) and L is the load factor, meaning the fraction of time over which the equipment is productively used.

The general form of the equation is revealed by assembling the terms into three groups:

$$\begin{array}{ccc} \text{[Material]} & \text{[Tooling]} & \text{[Time Capital Energy Space Information]} \\ \downarrow & \downarrow & \downarrow \quad \downarrow \quad \downarrow \quad \downarrow \quad \downarrow \end{array}$$

$$C = \left[\frac{mC_m}{1-f}\right] + \frac{1}{n}\left[C_t\right] + \frac{1}{\dot{n}}\left[\dot{C}_{L0} + \frac{C_c}{Lt_c} + \dot{P}C_e + A\dot{C}_S + \dot{C}_i\right] \tag{16.3}$$

The terms in the final bracket form a single 'gross overhead', $\dot{C}_{L,\text{gross}}$, allowing the equation to be written

$$\begin{array}{ccc} \text{[Materials]} & \text{[Dedicated cost/unit]} & \text{[Gross overhead/unit]} \\ \downarrow & \downarrow & \downarrow \end{array}$$

$$C = \left[\frac{mC_m}{1-f}\right] \quad + \quad \frac{1}{n}\left[C_t\right] \quad + \quad \frac{1}{\dot{n}}\left[\dot{C}_{L,\text{gross}}\right] \tag{16.4}$$

The equation indicates that the cost has three essential contributions: (1) a material cost/unit of production which is independent of production volume and rate, (2) a dedicated cost/unit of production which varies as the reciprocal of the production volume ($1/\dot{n}$), and (3) a gross overhead/unit of production which varies as the reciprocal of the production rate ($1/\dot{n}$). Plotted against the production volume, n, the cost, C, has the form shown in Figure 16.2. When the production volume, n, is small, the cost per kg of foam is totally dominated by the dedicated tooling costs C_t. As the production volume grows, the contribution of the second term in the cost equation diminishes. If the process is fast, the cost falls until, often, it flattens out at about twice that of the constituent materials.

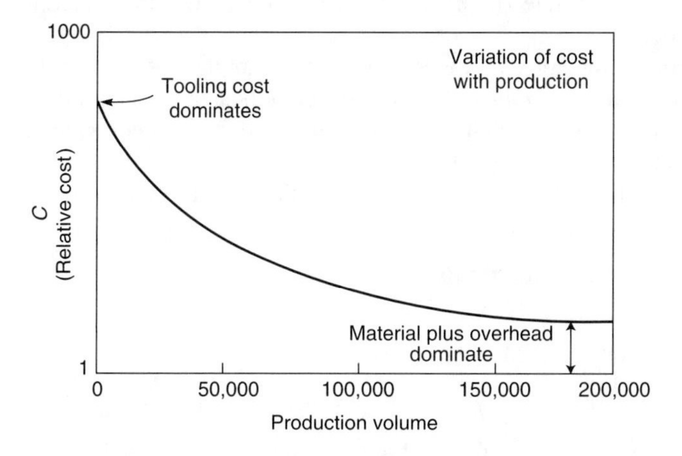

Figure 16.2 *The variation of material cost with production volume*

Technical cost modeling

Equation (16.3) is the first step in modeling cost. Greater predictive power is possible by introducing elements of *technical cost modeling* or TCM (Field and de Neufville, 1988; Clark *et al.*, 1997), which exploits the understanding of the way in which the control variables of the process influence production rate and product properties. It also uses information on the way the capital cost of equipment and tooling scale with output volume. These and other dependencies can be captured in theoretical and empirical formulae or look-up tables which are built into the cost model, giving greater resolution.

The elements of a TCM for liquid-state foaming of aluminum

Consider a cost model for the production of panels of a SiC-stabilized aluminum-based metallic foam by the process illustrated in Figure 16.3 and described in Chapter 2. There are four steps:

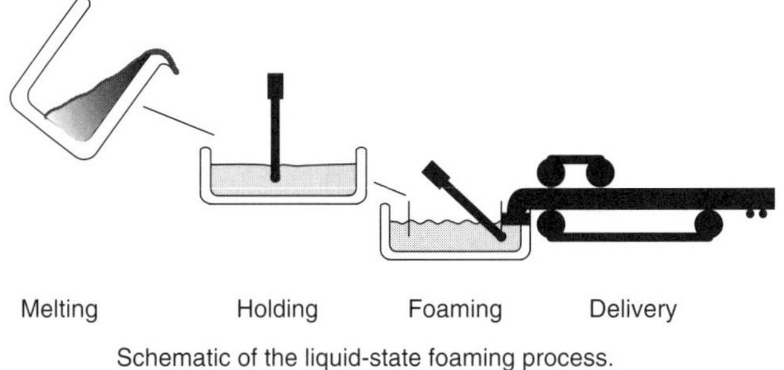

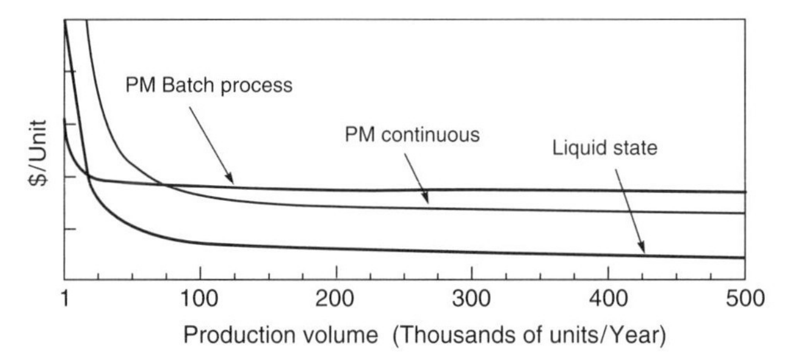

Figure 16.3 *Schematic of the liquid-state foaming process*

- Melting of the pre-mixed alloy
- Holding, providing a reservoir
- Foaming, using compressed gas and a bank of rotating blades
- Delivery, via a moving belt

The output of one step forms the input to the next, so the steps must match, dictating the size of the equipment or the number of parallel lines in each step. The capital and tooling costs, power, space and labor requirements for each step of the process are cataloged. Data are available characterizing the dependence of material density and of production rate on the gas flow rate and the stirring rate in the foaming step. These empirical relationships and relationships between equipment and production rate allow the influence of scale-up to be built into the model.

The outputs of the model (Figure 16.4) show the way in which the cost of the material depends of production volume and identifies cost drivers. Significantly, the model indicates the production volume which would be necessary to reach the plateau level of cost in which, in the best case shown here, the cost of the foam falls to roughly 2.5 times that of the constituent materials. Models of this sort (Maine and Ashby, 1999), applied to metal foam production, suggest that, with large-volume production, the cost of aluminum foams made by the melt foaming method (Section 2.2) could cost as little $3/lb; those made by the powder route, about $6/lb.

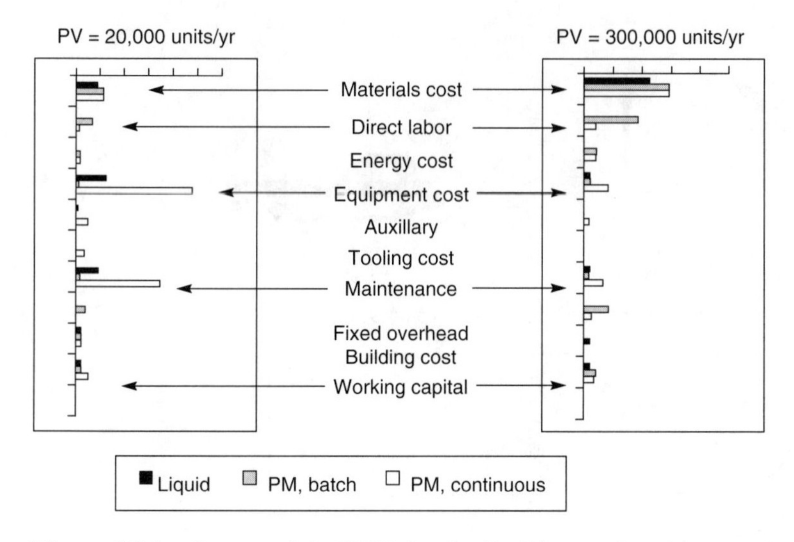

Figure 16.4 *Output of the TCM for the liquid-state foaming process of aluminum*

We now consider the final step in determining viability: that of value modeling.

16.4 Value modeling

Multi-objective optimization and trade-off surfaces

When a design has two or more objectives, solutions rarely exist that optimize all objectives simultaneously. The objectives are normally non-commensurate, meaning that they are measured in different units, and in conflict, meaning that any improvement in one is at the loss of another. The situation is illustrated for two objectives by Figure 16.5 in which one performance metric, P_2, is plotted another, P_1. It is usual to define the metrics such that a minimum is sought for each. Each bubble describes a solution. The solutions which minimize P_1 do not minimize P_2, and vice versa. Some solutions, such as that at A, are far from optimal: other solutions exist which have lower values of both P_1 and P_2, such as B. Solutions like A are said to be dominated by others. Solutions like that at B have the characteristic that no other solution exists with lower values of both P_1 and P_2. These are said to be non-dominated solutions. The line or surface on which they lie is called the non-dominated or optimum trade-off surface (Hansen and Druckstein, 1982; Sawaragi and Nakayama, 1985). The values of P_1 and P_2 corresponding to the non-dominated set of solutions are called the Pareto set.

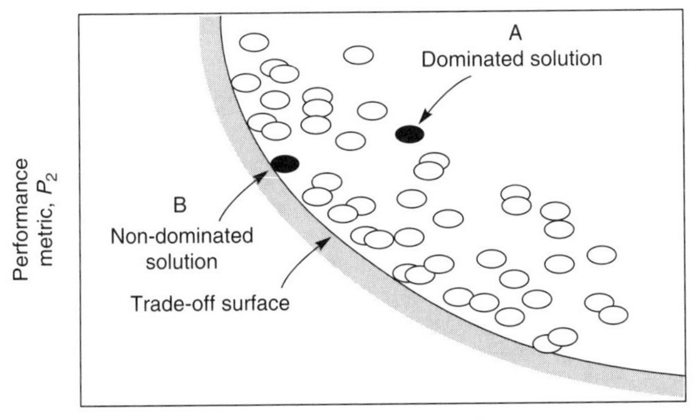

Figure 16.5 *Dominated and non-dominated solutions, and the optimum trade-off surface*

The trade-off surface identifies the subset of solutions that offer the best compromise between the objectives, but it does not distinguish between them. Three strategies are available to deal with this:

1. The trade-off surface, like that of Figure 16.5, is established and studied, using intuition to select between non-dominated solutions.
2. All but one of the objectives are reformulated as constraints by setting lower and upper limits for them, thereby allowing the solution which minimizes the remaining objective to be read off, as illustrated in Figure 16.6.

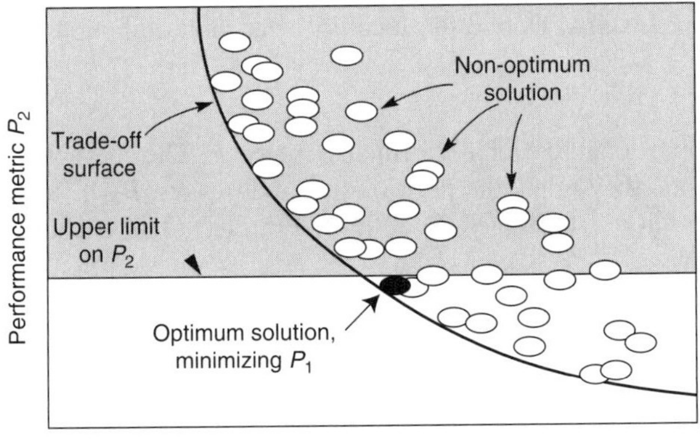

Figure 16.6 *Imposing limits on all but one of the performance metrics allows the optimization of the remaining one, but this defeats the purpose of multi-objective optimization*

3. A composite objective function or *value function*, V, is formulated; the solution with the minimum value of V is the overall optimum, as in Figure 16.7. This method allows true multi-objective optimization, but requires more information than the other two. It is explored next.

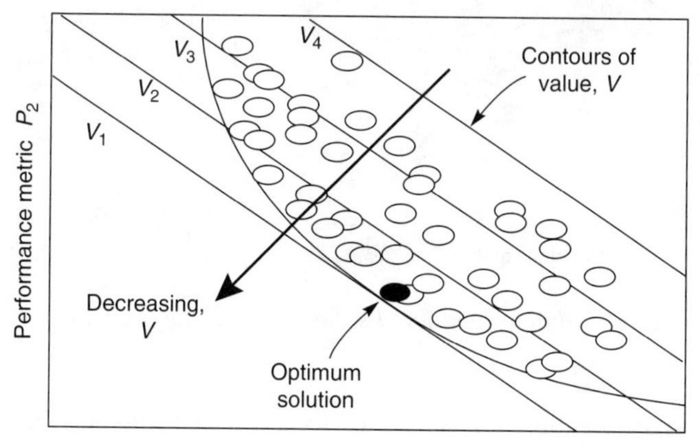

Figure 16.7 *A value function, V, plotted on the trade-off diagram. The solution with the lowest V is indicated. It lies at the point at which the value function is tangent to the trade-off surface*

Value functions

Value functions are used to compare and rank competing solutions to multi-objective optimization problems. Define the locally linear value function

$$V = \alpha_1 P_1 + \alpha_2 P_2 + \ldots\ldots \alpha_i P_i \ldots. \quad (16.5)$$

in which value V is proportional to each performance metric P. The coefficients α are exchange constants: they relate the performance metrics $P_1, P_2 \ldots$ to V, which is measured in units of currency (\$, £, DM, FF, etc.). The exchange constants are defined by

$$\alpha_1 = \left(\frac{\partial V}{\partial P_1}\right)_{P_2,\ldots P_i} \quad (16.6a)$$

$$\alpha_2 = \left(\frac{\partial V}{\partial P_2}\right)_{P_1,\ldots P_i} \quad (16.6b)$$

that is, they measure the change in V for a unit change in a given performance metric, all others held constant. If the performance metric P_1 is mass m (to be minimized), α_1 is the change in value V associated with unit increase in m. If the performance metric P_2 is heat transfer q per unit area, α_2 is the change in value V associated with unit increase in q. The best solution is the one with the smallest value of V, which, with properly chosen values of α_1 and α_2, now correctly balances the conflicting objectives.

With given values of V and exchange constants α_i, equation (16.5) defines a relationship between the performance metrics, P_i. In two dimensions, this plots as a family of parallel lines, as shown in Figure 16.7. The slope of the lines is fixed by the ratio of the exchange constants, α_1/α_2. The best solution is that at the point along a value-line that is tangent to the trade-off surface, because this is the one with the smallest value of V.

The method can be used to optimize the choice of material to fill a multi-functional role, provided that the exchange constants α_i for each performance metric are known.

Minimizing cost as an objective

Frequently one of the objectives is that of minimizing cost, C, so that $P_1 = C$. Since we have chosen to measure value in units of currency, unit change in C gives unit change in V, with the result that

$$\alpha_1 = \left(\frac{\partial V}{\partial P_1}\right)_{P_2,\ldots P_i} = 1 \tag{16.6c}$$

and equation (16.5) becomes

$$V = C + \alpha_1 P_2 + \ldots \alpha_i P_i \ldots. \tag{16.7}$$

As a simple example, consider the substitution of a metal foam, M, for an incumbent (non-foamed) material, M_0, based on cost, C, and one other performance metric, P. The value of P in the application is α. Substitution is potentially possible if the value V of M is less than that, V_0, of the incumbent M_0. Thus substitution becomes a possibility when

$$V - V_0 = (C - C_0) + \alpha\,(P - P_0) \leq 0 \tag{16.8}$$

or

$$\Delta V = \Delta C + \alpha \Delta P \leq 0$$

from which

$$\frac{\Delta P}{\Delta C} \leq -\frac{1}{\alpha} \tag{16.9}$$

defining a set of potential applications for which M is a better choice than M_0.

To visualize this, think of a plot of the performance metric, P, against cost, C, as shown in Figure 16.8 The incumbent M_0 is centered at $\{P_0, C_0\}$;

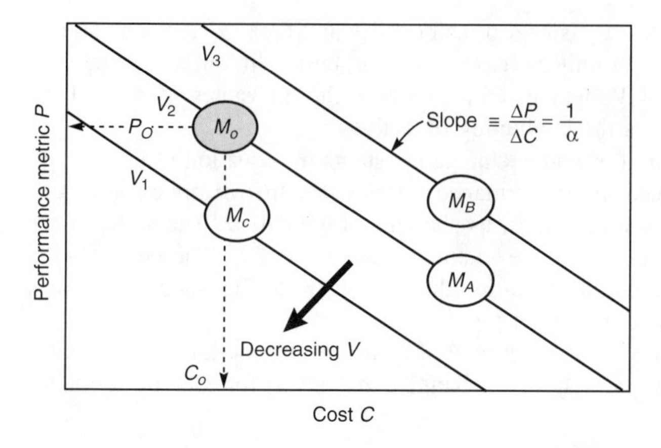

Figure 16.8 *The trade-off between performance and cost. Neither material M_A nor M_B is a viable substitute for M_0. Material M_C is viable because it has a lower value of V*

the potential substitute at $\{P, C\}$. The line through M_0 has the slope defined by equation (16.9), using the equality sign. Any material which lies on this line, such as M_A, has the same value V as M_0; for this material, ΔV is zero. Materials above this line, such as M_B, despite having a lower (and thus better) value of P than M_0, have a higher value of V. Materials below the line, such as M_C, have a lower value of V (even though they cost more), a necessary condition for substitution. Remember, however, that while negative ΔV is a necessary condition for substitution, it may not be a sufficient one; sufficiency requires that the difference in value ΔV be large enough to justify the investment in new technology.

Values for the exchange constants α_i

An exchange constant is a measure of the value, real or perceived, of a performance metric. Its magnitude and sign depend on the application. Thus the value of weight saving in a family car is small, though significant; in aerospace it is much larger. The value of heat transfer in house insulation is directly related to the cost of the energy used to heat the house; that in a heat exchanger for power electronics can be much higher. The value of performance can be real, meaning that it measures a true saving of cost, energy, materials, time or information. But value can, sometimes, be perceived, meaning that the consumer, influenced by scarcity, advertising or fashion, will pay more or less than the true value of these metrics.

In many engineering applications the exchange constants can be derived approximately from technical models. Thus the value of weight saving in transport systems is derived from the value of the fuel saved or that of the increased payload that this allows (Table 16.2). The value of heat transfer can be derived from the value of the energy transmitted or saved by unit change in the heat flux per unit area. Approximate exchange constants can sometimes be derived from historical pricing data; thus the value of weight saving in bicycles can be found by plotting the price* P of bicycles against their mass m, using the slope dP/dm as a measure of α. Finally, exchange constants can be found by interviewing techniques (Field and de Neufville, 1988; Clark *et al.*, 1997), which elicit the value to the consumer of a change in one performance metric, all others held constant.

Table 16.2 *Exchange constants α for transport systems*

Sector: Transport systems	*Basis of estimate*	*Exchange constant £/kg ($/lb)*
Car, structural components	Fuel saving	0.5 to 1.5 (0.4 to 1.1)
Truck, structural components	Payload, fuel saving	5 to 10 (4 to 8)
Civil aircraft, structural	Payload	100 to 500 (75 to 300)
Military vehicle, structural	Payload, performance	500 to 1000 (350 to 750)
Space vehicle, structural	Payload	3000 to 10 000 (2000 to 75 000)
Bicycle, structural components	Perceived value (derived from data for price and mass of bicycles)	80 to 1000 (50 to 700)

The values of α in Table 16.2 describe simple trade-offs between cost and performance. Circumstances can change these, sometimes dramatically. The auto-maker whose vehicles fail to meet legislated requirements for fleet fuel consumption will assign a higher value to weight saving than that shown in Table 16.2; so, too, will the aero-engine maker who has contracted to sell an engine with a given power-to-weight ratio if the engine is overweight. These special cases are not uncommon, and can provide the first market opportunity for a new material.

* For any successful product the cost, C, the price, P, and the value, V, are related by $C < P < V$, since if $C > P$ the product is unprofitable, and if $P > V$ no one will buy it. Thus P can be viewed as a lower limit for V.

16.5 Applications

Two examples of the method follow, each exploring the viability of metal foams in a particular application. In the first, metal foams prove to be non-viable. In the second, despite their present high cost, they prove to be viable. The examples are deliberately simplified to bring out the method. The principles remain the same when further detail is added.

Simple trade-off between two performance indices

Consider selection of a material for a design in which it is desired, for reasons of vibration control, to maximize the specific modulus E/ρ (E is Young's modulus and ρ is the density) and the damping, measured by the loss coefficient η. We identify two performance metrics, P_1 and P_2, defined such that minima are sought for both:

$$P_1 = \frac{\rho}{E} \tag{16.10a}$$

and

$$P_2 = \frac{1}{\eta} \tag{16.10b}$$

Figure 16.9 shows the trade-off plot. Each bubble on the figure describes a material; the dimensions of the bubble show the ranges spanned by these property groups for each material. Materials with high values of P_1 have low values of P_2, and vice versa, so a compromise must be sought. The optimum trade-off surface, suggested by the shaded band, identifies a subset of materials with good values of both performance metrics. If high E/ρ (low P_1) is of predominant importance, then aluminum and titanium alloys are a good choice; if greater damping (lower P_2) is required, magnesium alloys or cast irons are a better choice; and if high damping is the over-riding concern, tin or lead alloys and a range of polymers become attractive candidates. It is sometimes possible to use judgement to identify the best position on the trade-off surface (strategy 1, above). Alternatively (strategy 2) a limit can be set for one metric, allowing an optimum for the other to be read off. Setting a limit of $\eta > 0.1$, meaning $P_2 < 10$, immediately identifies pure lead and polyethylenes as the best choices in Figure 16.9. Finally, and preferably (strategy 3), a value function can be determined:

$$V = \alpha_1 P_1 + \alpha_2 P_2 = \alpha_1 \frac{\rho}{E} + \alpha_2 \frac{1}{\eta} \tag{16.11}$$

seeking materials which minimize V. Contours of constant V, like those of Figure 16.7, have slope

$$\left(\frac{\partial P_2}{\partial P_1}\right)_V = -\frac{\alpha_1}{\alpha_2} \tag{16.12}$$

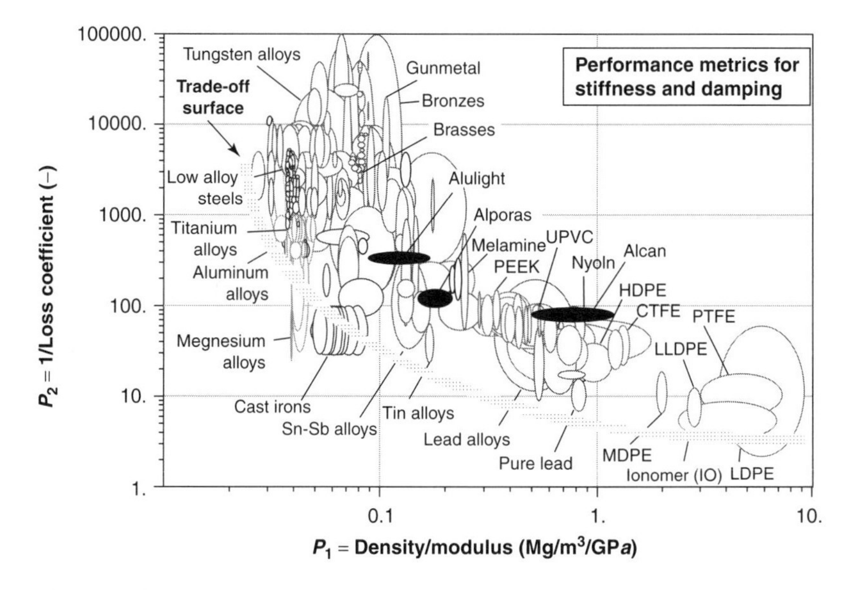

Figure 16.9 *A trade-off plot for the performance metrics $P_1 = \rho/E$ and $P_2 = 1/\eta$. Each bubble refers to a material class. The metal foams are distinguished by filled ellipses (all other materials are fully dense). The shaded band show the optimum trade-off surface. Materials that lie on or near this surface most nearly optimize both performance metrics*

The point at which one contour is tangent to the trade-off surface identifies the best choice of material. Implementation of this strategy requires values for the ratio α_1/α_2 which measures the relative importance of stiffness and damping in suppressing vibration. This requires estimates of the influence of each on overall performance, and can, in technical systems, be modeled. Here, however, it is unnecessary. The positions of three classes of metal foams are shown as black ovals. None lie on or near the trade-off surface; all are sub-dominant solutions to this particular problem. Metal foams, in this application, are non-viable.

Co-minimizing mass and cost

One of the commonest trade-offs is that between mass and cost. Consider, as an example, co-minimizing the mass and cost of the panel of specified bending stiffness analysed in Chapter 5, Section 5.3. The mass of the panel is given by equation (5.4) which we rearrange to define the performance metric P_1:

$$P_1 = \frac{m}{\beta} = \left(\frac{\rho}{E^{1/3}}\right) \tag{16.13}$$

with β given by

$$\beta = \left(\frac{12S^* b^2}{C_1}\right)^{1/3} \ell^2 \tag{16.14}$$

β is a constant for a given design, and does not influence the optimization. The geometric constant, C_1, defined in equation (5.4), depends only on the distribution of loads on the panel and does not influence the subsequent argument. The cost, C_p, of the panel is simply the material cost per kg, C (from equation (16.4)), times the mass m, giving the second performance metric P_2:

$$P_2 = \frac{C_p}{\beta} = \left(\frac{C\rho}{E^{1/3}}\right) \tag{16.15}$$

Figure 16.10 shows the trade-off plot. The horizontal axis, P_1, is the material index $\rho/E^{1/3}$. The vertical axis, correspondingly, is the index $C\rho/E^{1/3}$.

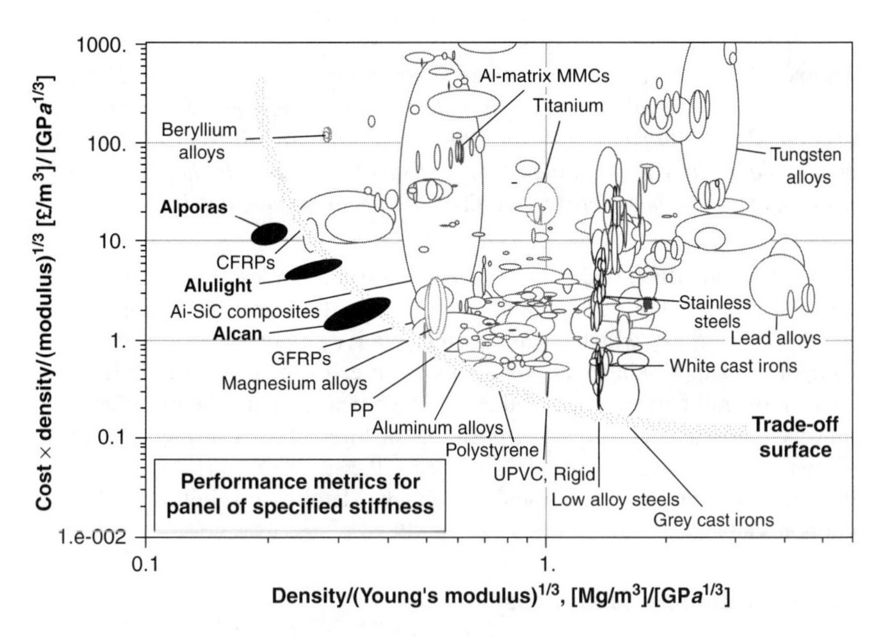

Figure 16.10 *A trade-off plot for the performance metrics (measured by material indices) for cost and mass of a panel of specified bending stiffness. Each bubble refers to a material class. The metal foams are distinguished by filled ellipses (all other materials are fully dense). The trade-off front, constructed for non-foamed materials, separates the populated section of the figure from that which is unpopulated. Metal foams lie in the unpopulated sector*

Conventional alloys (cast irons, steels, aluminum alloys) lie in the lower part of the diagram. Beryllium alloys, CFRPs and Al-based MMCs lie in the central and upper parts.

The trade-off surface for conventional, fully dense, materials is shown by the shaded band. Metal foams lie in the unpopulated sector of the diagram – all three classes of foam offer non-dominated solutions to this problem. But even so, they are viable only if the mass/value exchange constant lies in the right range. To explore this question for the panel of specified stiffness we define the value function

$$V = \alpha_1 P_1 + \alpha_2 P_2 = \alpha_1 \left(\frac{\rho}{E^{1/3}}\right) + \left(\frac{C\rho}{E^{1/3}}\right) \tag{16.16}$$

(since α_2, relating value to cost, is unity). Values of α_1, relating value to mass, are listed in Table 16.2. The equation is evaluated in Table 16.3 for two extreme values of α_1 for a set of materials including cast irons, steels, aluminum alloys, titanium, beryllium and three metal foams. When α_1 has the low value of 0.5 £/kg, nodular cast irons are the best choice. But if α_1 is as

Table 16.3 *The selection of panel materials: stiffness constraint*

Material	ρ *Mg/m³*	*E* *GPa*	C_m *£/kg*	P_l	P_2	*V* $\alpha_l =$ *£0.5/kg*	*V* $\alpha_l =$ *£500/kg*
Cast iron, nodular	7.30	175	0.25	1.31	0.33	0.99	655
Low-alloy steel (4340)	7.85	210	0.45	1.32	0.59	1.25	660
Al-6061-T6	2.85	70	0.95	0.69	0.66	1.01	345
Al-061–20% SiC, PM	2.77	102	25	0.59	14.8	15.1	309
Ti-6-4 B265 grade 5	4.43	115	20	0.91	18.2	18.7	473
Beryllium SR-200	1.84	305	250	0.27	67.5	67.6	202
Alporas[a]	0.25	1.0	40	0.23	10.0	10.1	125
Alulight[a]	0.30	0.8	16	0.3	5.2	5.4	155
Alcan[a]	0.25	0.4	5.8	0.34	2.0	2.2	172

[a] All three types of metal foam are made in a range of densities, with a corresponding range of properties. These three examples are taken from the middle of the ranges. The costs are estimates only, broadly typical of current prices, but certain to change in the future. It is anticipated that large-scale production could lead to substantially lower costs.

high as 500 £/kg, all three of the foams offer better value V than any of the competing materials. Alporas, using present data, is the best choice, meaning that it has the best combination of performance and cost.

Multifunctionality

Metal foams appear to be most attractive when used in a multifunctional role. Multifunctionality exploits combinations of the potential applications listed in Table 1.2. The most promising of these are

- Reduced mass
- Energy absorption/blast mitigation
- Acoustic damping
- Thermal stand-off (firewalls)
- Low-cost assembly of large structures (exploiting low weight)
- Strain isolation (as a buffer between a stiff structure and a fluctuating temperature field, for instance).

The method described above allows optimization of material choice for any combination of these.

References

Ashby, M.F. (1997) *Materials Selection: Multiple Constraints and Compound Objectives*, American Society for Testing and Materials, STP 1311, pp. 45–62.

Ashby, M.F. (1999) *Materials Selection and Mechanical Design*, 2nd edition, Butterworth-Heinemann, Oxford.

Ashby, M.F. (2000) Multi-objective optimisation in material design and selection. To appear in *Acta Mater*, January 2000.

Bader, M.G. (1997) Proc. ICCM-11, Gold Coast, Australia, Vol. 1: *Composites Applications and Design*, ICCM, London, UK.

Clark, J.P., Roth, R. and Field, F.R. (1997) Techno-economic issues in material science. In ASM Handbook Vol. 20, *Materials Selection and Design*, ASM International, Materials Park, OH, 44073-0002, USA.

Esawi, A.M.K. and Ashby, M.F. (1998) Computer-based selection of manufacturing processes: methods, software and case studies. *Proc. Inst. Mech. Engrs* **212B**, 595–610.

Field, F.R. and de Neufville, R. (1988) Material selection – maximising overall utility. *Metals and Materials* June, 378.

Hansen, D.R. and Druckstein, L. (1982) *Multi-objective Decision Analysis with Engineering and Business Applications*, Wiley, New York.

Maine, E.M.A. and Ashby, M.F. (1999) Cost estimation and the viability of metal foams. In Banhart, J., Ashby, M.F. and Fleck, N.A. (eds), *Metal Foams and Foam Metal Structures* Proc. Int. Conf. Metfoam '99, 14–16 June 1999, MIT Verlag, Bremen, Germany.

Sawaragi, Y. and Nakayama, H. (1985) *Theory of Multi-Objective Optimisation*, Academic Press, New York.

Chapter 17

Case studies

Metal foams already have a number of established and profitable market niches. At the high value-added end of the market, the DUOCEL® range of metfoams are used as heat exchangers, both of the regenerative and the purely dissipative types; they are used, too, as lightweight support structures for aerospace applications. The INCO® nickel foams allow efficient, high current-drain rechargeable batteries. ALPORAS® aluminum foams have been deployed as baffles to absorb traffic noise on underpasses and as cladding on buildings; CYMAT® aluminum foams can be used both for lightweight cladding and as structural panels.

These are established applications. Potential applications under scrutiny include automobile firewalls, exploiting thermal and acoustic properties as well as mechanical stiffness and energy-absorbing ability at low weight; and components for rail-transport systems, exploiting the same groups of properties. Other potential applications include integrally molded parts such as pump housings and hydraulic components. Some metfoams are potentially very cheap, particularly when cost is measured in units of $/unit volume. Here they rival wood (without the problems of decay) and other cheap structural materials such as foamed plastics.

The eight case studies together with the potential applications listed in this chapter and in Chapter 1, Table 1.2, will give an idea of the possibilities for exploiting metal foams in engineering structures.

17.1 Aluminum foam car body structures

Karmann GmbH is a system supplier for the automotive industry worldwide.* The company designs and produces vehicles for original equipment

* Contact details: W. Seeliger, Wilhelm Karmann GmbH, Karmann Strasse 1, D-49084, Osnabrück., Germany. Phone (+49) 5 41-581-0; fax (+49) 5 41-581-1900.

Karmann USA Inc., 17197 North Laurel Park Drive, Suite 537, Livonia, Michigan 48152, USA Phone (313) 542-0106; fax (313) 542-0305.

Schunk Sintermetalltechnik GmbH, Postfach 10 09 51, D-35339 Gießen, Germany.

Yu, M and Banhart, J. (1998) In *Metal Foams*, (1998) Proc. Fraunhofer USA Metal Foam Symposium, Stanton, NJ, MIT Verlag, Bremen, Germany.

makers (OEMs). Karmann has announced a newly developed Aluminum Foam System allowing revolutionary technology in body panels (Figures 17.1 and 17.2). It claims that aluminum foams offer cost-effective performance as structural automotive parts that are up to ten times stiffer and 50% lighter than equivalent parts made of steel. Such lightweight, stiff foam sandwich panels simplify body structure systems, enabling OEMs to produce different variations of low-volume niche vehicles based on a common body structure.

Figure 17.1 *A concept design for a low-weight vehicle. The firewall and trunk are made of three-dimensional aluminum foam panels. (Courtesy of Karmann GmbH)*

The three-dimensional aluminum foam system consists of outer skins roll-bonded to a central layer containing a dispersion of titanium hydride (see Chapter 2, Section 2.4). The central section is expanded by heat treatment after the panel has been pressed to shape. Sections of the body shell considered well suited for aluminum foam sandwich panels include firewalls, roof panels and luggage compartment walls.

As much as 20% of the auto structure could be made from three-dimensional aluminum foam panels. The company note that, in the typical compact family sedan, this would lead to a mass saving of 60 kg, translating into a reduction in fuel consumption of 2.6 miles per gallon.

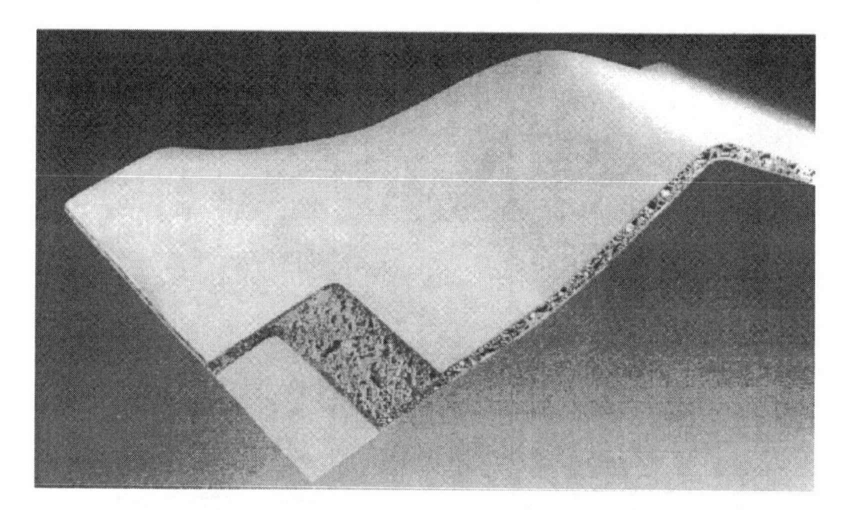

Figure 17.2 *A pressed panel after expansion, showing the three-layer structure. (Courtesy of Karmann GmbH)*

17.2 Integrally molded foam parts

Illichmann GmbH give examples of the ways in which MEPURA ALULIGHT® can be molded to give complex shapes which have dense skins with foamed cores – 'syntactic' metal structures* (see Chapter 2, Section 2.4). Figure 17.3 shows examples of skinned structures which are both stiff and light. The thickness of the outer skin can be enlarged by a special casting technique, creating three-dimensional sandwiches with isotropic core properties. The small amount of metal ($<40\%$) reduces the thermal conductivity significantly, but the electrical conductivity remains high. The structure has good damage tolerance and energy-absorbing capability.

The moldings are made from various aluminum alloys, primarily (1) commercially pure aluminum; (2) heat-treatable aluminum alloys of the 6000-series and (3) aluminum alloys based on AlSi12 casting alloy. Mechanical properties can be optimized by appropriate heat treatment of the base alloy. Surface treatment is also possible using standard techniques for aluminum and its alloys.

* Contact details: Illichmann GmbH Grossalmstrasse 5, A-4813, Altmünster, Austria. Phone: (+43) 761288055-0; fax (+43) 761288055-29.

Mepura Metallpulvergesellschaft m.b.H., Ranshofen, A-4813 Altmünster Grossalmstrasse, 5, Austria. Phone: (+43) 761288055-0; Fax (+43) 772268154.

Dr Frantesek Simancik, Institute of Materials and Machine Mechanics, Slovak Academy of Sciences, Racianska 75, PO Box 95, 830 08 Bratislava 38, Slovak Republic. Phone: (+42) 7254751; fax (+42) 7253301.

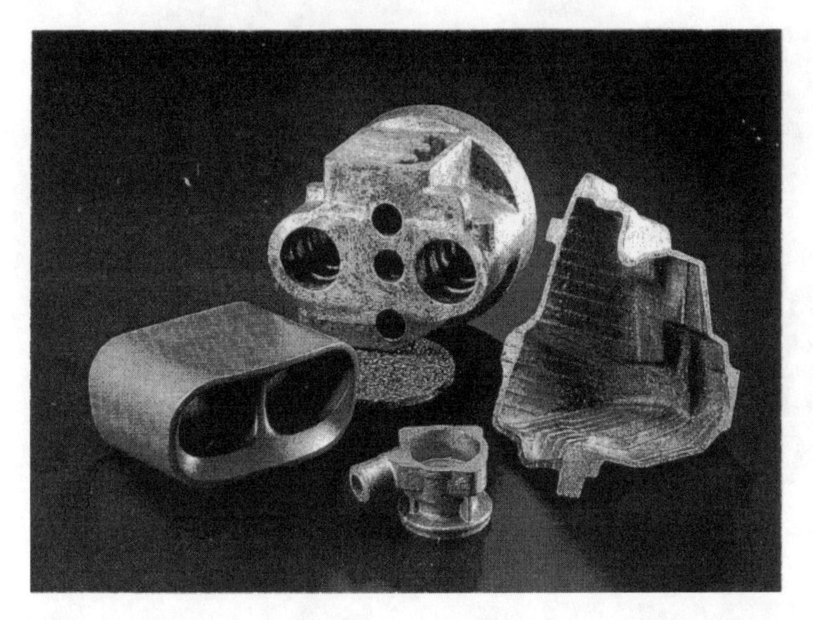

Figure 17.3 *Foamed aluminum components with integral skins. (Courtesy of Mepura Metallpulvergesellschaft m.b.H)*

MEPURA and Illichmann suggest applications for ALULIGHT moldings which include

- Lightweight machine castings with improved sound and vibration damping
- Impact energy absorption components for cars, lifting and conveyor systems
- Stiff machine parts with significantly reduced weight
- Housings for electronic devices providing electromagnet and thermal shielding
- Permanent cores for castings, replacing sand cores
- Isotropic cores for sandwich panels and shells
- Fillings in hollow shapes to inhibit buckling
- Heat shields and encapsultors
- Floating structures at elevated temperatures and pressures
- Sound absorbers for difficult conditions

17.3 Motorway sound insulation

Shinko Wire Company Ltd has developed ALPORAS® for use as sound-proofing material that can be installed along the sides of a road or highway

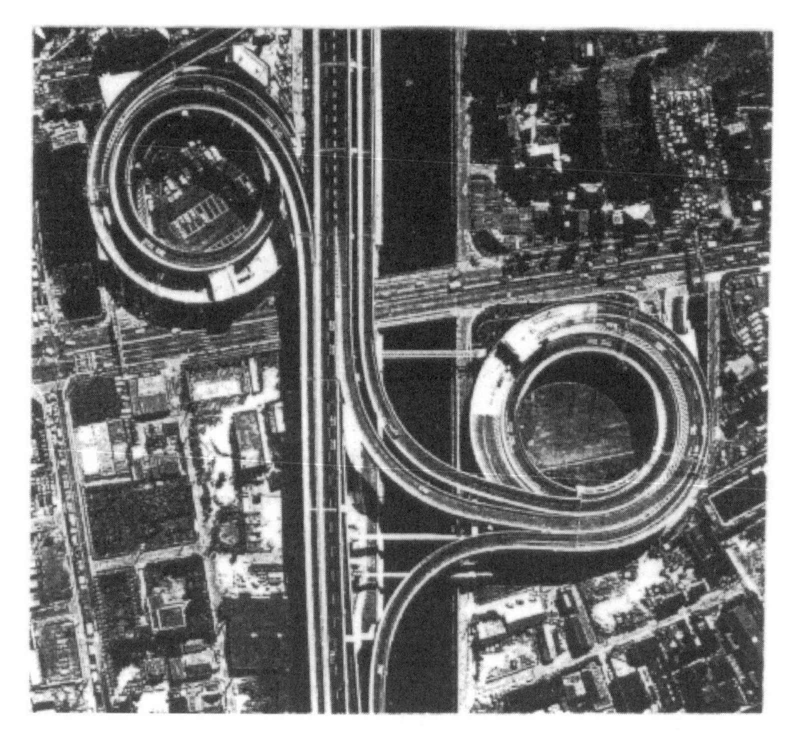

Figure 17.4 *Highways, high traffic volume and high residential development create the need for sound-management systems. (Courtesy of Shinko Wire Company Ltd.)*

to reduce traffic noise.* With the development of extensive highways and the increase in traffic volume and the density of residential development, noise pollution and other environmental problems have assumed greater prominence, creating the need for sound shielding. A proprietary treatment of ALPORAS (see Chapter 2, Section 2.3) gives a foam structure with enhanced sound-absorbing capability. The structure features a layer of foamed aluminum that is attached to either concrete or galvanized steel with an air gap of calculated width to maximize absorption. The foamed aluminum faces the road to maximize sound absorption. The concrete or galvanized steel backing acts as a sound insulator, keeping the noise from reaching surrounding residents.

Shinko claim that the material is fire resistant, does not generate harmful gases in the presence of a flame, has excellent durability and resistance to

* Shinko Wire Company Ltd, Alporas Division, 10-1 Nakahama, Amagasaki 660, Japan. Phone: (06) 411-1061; fax (06) 411-1075.

Drillfix AG, Drillfix European Distribution, Herrenmatt 7F, CH-5200 Brugg, Switerland. Phone (41) 56 442 5037; fax (41) 564423635.

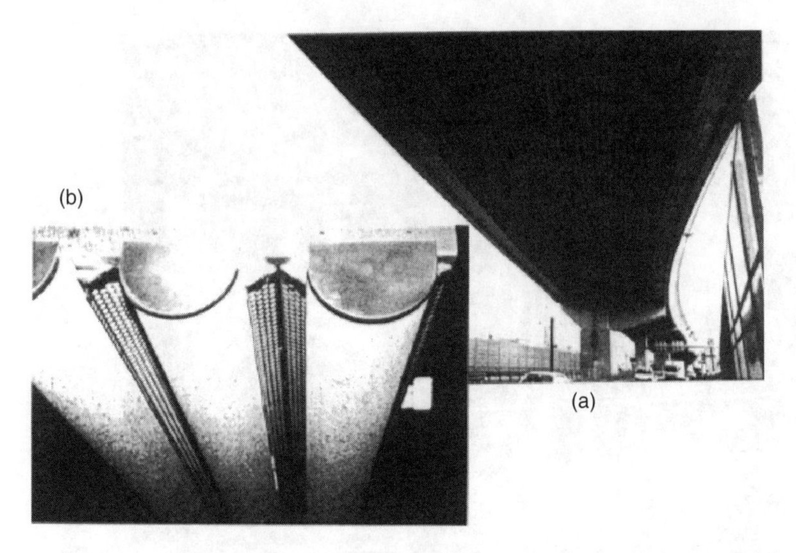

Figure 17.5 *(a) Sound-absorbing lining on the underside of a highway bridge; (b) the sound-absorbing elements are shaped like hemi-circular tubes*

weathering, does not absorb water, and can be washed down to keep it clean. The sound-absorbing structures have some shock-absorbing capacity, an attractive appearance, and act as electromagnetic shields, limiting ignition and other electromagnetic disturbances from passing vehicles, and shock waves caused by tunnel sonic boom from ultra-high-speed trains.

17.4 Optical systems for space applications

ERG, the producers of the DUOCEL® range of metal foams (see Chapter 2, Section 2.5), exemplify successful applications of these materials in aerospace.* This case study and the next two describe three of these.

Mirrors of all sorts play a key role in space technology. Whether the frequencies are optical, infrared or microwave, precision, low mass and freedom from long-term distortion are primary design constraints. Figure 17.6 shows the basic structure of a DUOCEL-backed composite mirror used by both the Lockheed Missile and Space Co. and the Hughes Aircraft Company Laser Systems Division. The mirror face is stiffened by bonding it to a foam layer backed by a stiff backing plate. The structure is light and very stiff; the all-metal construction gives freedom from long-term distortion.

Optical (and other) systems for space application must retain their precision despite extreme changes in levels of solar energy exposure. Figure 17.7

* Contact details: ERG Materials and Aerospace Corporation, 900 Stanford Ave, Oakland, CA 94608, USA. Phone: (510) 658-9785; Fax: (510) 658-7428.

Figure 17.6 *DUOCEL® foamed aluminum used as the structural core of a lightweight composite mirror. (Courtesy of ERG)*

Figure 17.7 *DUOCEL® foamed aluminum sunshade for an optical telescope. (Courtesy of ERG)*

illustrates a DUOCEL foamed aluminum composite sunshade for an optical telescope satellite.

17.5 Fluid–fluid heat exchangers

Figure 17.8 illustrates the structure of a fluid–fluid heat exchanger constructed from ERG DUOCEL® open-celled aluminum foam.* Heat exchange criteria are analysed in Chapter 13 of this Design Guide. Open porosity, low relative

Figure 17.8 *DUOCEL® foamed aluminum used as the heat-exchange medium for the space shuttle atmospheric control system. (Courtesy of ERG)*

* Contact details: ERG Materials and Aerospace Corporation, 900 Stanford Ave, Oakland, CA 94608, USA. Phone: (510) 658-9785; Fax: (510) 658-7428.

density and high thermal conductivity of the cell edges are essential. The DUOCEL range of metal foams, which include both aluminum and copper, offer this combination of properties. The figure shows a foamed aluminum heat exchanger for a space-shuttle atmospheric control system.

17.6 Light weight conformal pressure tanks

Fluids in pressure tanks, ranging from aircraft fuel to liquid nitrogen cryogen for airborne infrared telescopes, must be maintained at constant and uniform temperature, uniform pressure, and – in moving systems – must be restrained from 'sloshing' because this generates pressure gradients. Figure 17.9 shows the design of a tank which exploits ERG DUOCEL® open-celled aluminum foam to achieve this. The foam has a low density, occupying only 10% of the volume of the tank, but it limits fluid motion. The high thermal conductivity of the cell edges maintains the fluid at a constant temperature, important for jet fuel.*

Figure 17.9 *DUOCEL® foamed aluminum used as the structural core, heat exchanger and anti-slosh baffle in a lightweight conformal tank. (Courtesy of ERG)*

17.7 Electrodes for batteries

Nickel rechargeable batteries systems, nickel cadmium (NiCd) and nickel metal hydride (NiMH) are the most widely used batteries for consumer

* Contact details: ERG Materials and Aerospace Corporation, 900 Stanford Ave, Oakland, CA 94608, USA. Phone: (510) 658-9785; Fax:(510) 658-7428.

portable applications such as power tools, video cameras and cellular phones.* Different nickel battery designs are required to meet these various applications. Properties can be varied to meet the needs of power, cycle life and energy density.

When energy density is the most important characteristic, nickel batteries have made gains by using nickel foam as the electrodes, and the development of the nickel metal hydride system with a hydrogen-absorbing nickel alloy as the negative electrode. Nickel foams can be made to different densities, thicknesses and porosity to optimize performance using the process described in Chapter 2, Section 2.6.

These nickel battery systems are now being specified by the world's leading automotive companies (GM, Toyota, Honda) for the next generation of electric and hybrid vehicles. Nickel foams can also be used as filtration media.

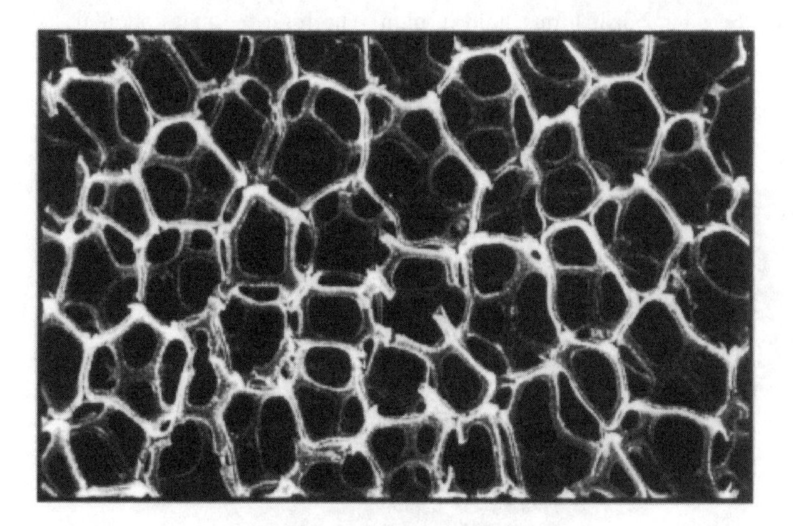

Figure 17.10 *A nickel foam positive electrode as used in NiCd and NiMeH batteries. (Courtesy of Inco Ltd.)*

17.8 Integrated gate bipolar transistors (IGBTs) for motor drives

Present motor drives generally comprise integrated gate bipolar transistors (IGBTs) because they are capable of operating at the high power densities

* Contact details: Inco Ltd, 145 King Street West, Suite 1500, Toronto, Canada, M5H 4B7. Phone: (416) 361-7537; Fax: (416) 361-7659.

Inco European Ltd, London office, 5th floor, Windsor House, 50 Victoria Street, London SW1H OXB. Phone (44) 171 932-1516; fax (44) 171 9310175.

required for compactness. Each IGBT in commercial systems is configured as shown in Figure 17.11. Each module comprises several IGBTs with an equal number of diodes (six would be typical for a 75 hp motor drive). In steady operation, the heat, q, generated at the electronics in each IGBT may be as large as $6\,\mathrm{MW/m^2}$. The flux is in one direction and is transferred to the coolant by a heat sink comprising a fin-pin array subject to flowing air generated by a fan. Conventional air-cooled heat sinks operate at fluxes, $\overline{q} \leqslant 2\,\mathrm{kW/m^2}$ (Figure 17.12). The ratio $\overline{q}/q$ is accommodated by designing the sink with a cross-sectional dimension, b_{hs}, that relates to that for the Si, b_{si},

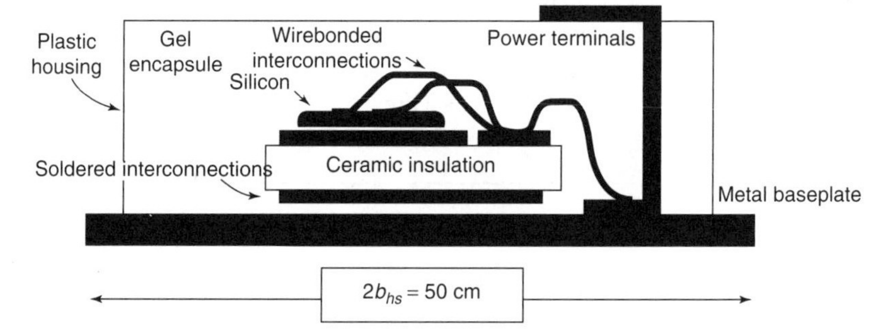

Figure 17.11 *Conventional power electronic packaging*

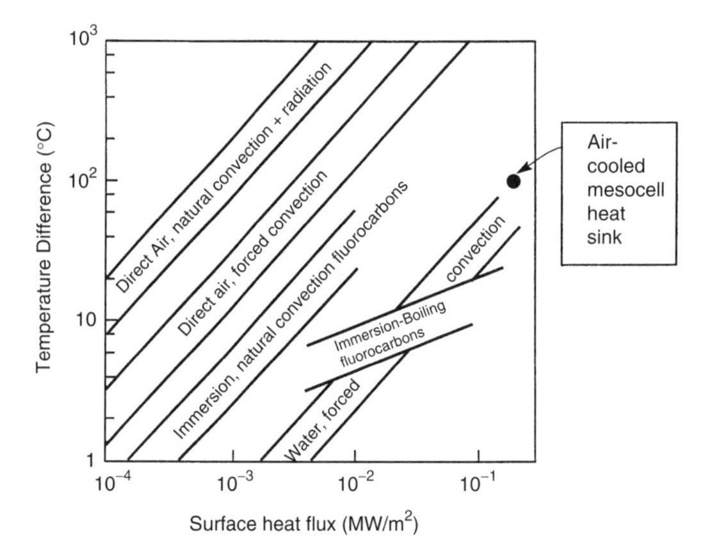

Figure 17.12 *Performance domains for heat sinks*

in accordance with:

$$\overline{q}/q = (b_{si}/b_{hs})^2 \tag{17.1}$$

With $b_{\rm si} = \frac{1}{2}$ cm, this requirement results in $b_{\rm hs} = 25$ cm, causing the overall system to occupy a large volume. The goal of the case study is to reduce b_{hs} to 3 cm, by using a cellular metallic heat sink, resulting in an order of magnitude reduction in overall volume.

The power density at the electronics is limited by the temperature reached at the junction, T_j. For Si electronics, T_j must be less than 120°C to avert unacceptable degradation. The design of the system and the importance of the heat transfer coefficient at the sink interrelate through $\overline{q}$, q and T_j. The capacity of the fluid pumping system is another key factor. Such systems are characterized by an operating curve that connects the back pressure to the allowable fluid flow rate through the sink (Figure 17.13).

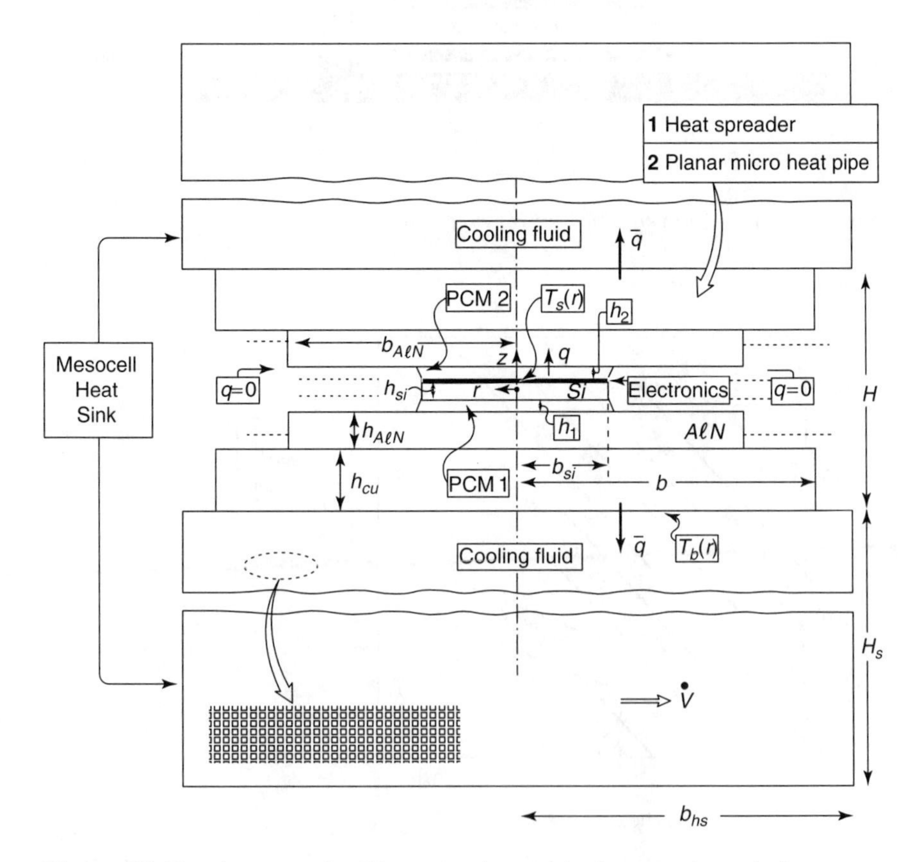

Figure 17.13 *An example of integrated gate bipolar transistor design*

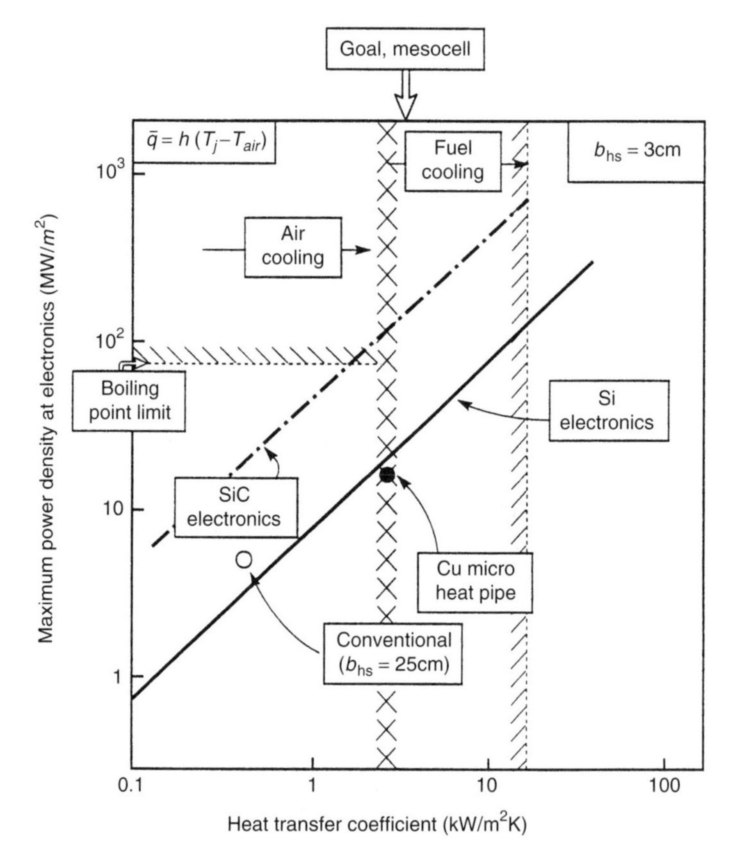

Figure 17.14 *Upper limits on power density and heat-transfer coefficient*

This case study illustrates the benefits of using a cellular metal sink to cool power electronic devices. Analytical results provide upper bounds. Numerical simulations provide explicit operating benefits. The overall goal is to reduce the volume of the drive needed to operate, say, a 75 hp motor (relative to conventional IGBT modules) while increasing its durability and decreasing its cost.

Requirements

If two-sided cooling is used to double the heat flux achievable at the electronics, there is a maximum achievable flux at the sink, $\bar{q}^*$. This occurs when the device located between the electronics and the surface of the sink is designed to be nearly isothermal and equal to the junction temperature, T_j,

$$\bar{q}^* = \bar{h}(T_j - \bar{T}_f) \tag{17.2}$$

where $\overline{h}$ is the average heat transfer coefficient from the electronics into the coolant, and $\overline{T}_f$ is the average temperature of the fluid. The design used to approach this maximum entails the use of heat spreaders and planar micro-heat-pipes. Based on equations (17.1) and (17.2), for the heat sink to be reduced to 3 cm, a heat transfer coefficient exceeding about 3 kW/m^{2}K is needed. Such levels can be readily realized using liquids, but are well in excess of those now achievable with air cooling (Figure 17.12). The challenge is to determine whether cellular metals can attain such high $\overline{h}$ with air.

Cellular metal performance

For the airflows needed to realize these $\overline{h}$, a fan/blower configuration with requisite operating characteristics must be designed. These characteristics typically exhibit a nearly linear interdependence between back pressure Δp and volume flow rate, $\dot{V}$ (Table 17.1), with Δp^* and $\dot{V}^*$ as the respective perfor-

Table 17.1 *Formulae for calculating achievable heat dissipation*

(1) Heat extracted over sink area ($4b^2$)

$$\overline{q} = v[T_j - T_0]\Omega_f c_p (H_s/2b_{hs})[1 - \exp(-2b_{hs}/\ell)]$$

(2) Transfer length

$$\ell = \frac{4.1 v \Omega_f c_p H_s d}{(1-\rho) k_m Bi} \left\{ 1 + \frac{2\rho}{(1-\rho)\sqrt{Bi}} \tanh\left[\frac{H_s\sqrt{Bi}}{1.24d}\right]\right\}^{-1}$$

(3) Biot number

$$Bi = 1.04 \left[\frac{\Omega_f c_p}{k_f}\right]^{0.4} \left[\frac{vd}{1-\alpha}\right]^{0.4} \left(\frac{k_f}{k_m}\right)$$

(where $\alpha = 0.37\sqrt{\rho} + 0.055\rho$)

(4) Pressure drop in sink

$$\Delta p = 0.75\sqrt{\rho}\left[\frac{\Omega_f^{0.4} v^{1.6}}{d^{1.4}(1-\alpha)^{1.6}}\right]$$

(5) Operating characteristics of pump

$$\Delta p = \Delta p^*[1 - v h_s b_{hs}/\dot{V}^*]$$

(6) Fluid flow rate

$$\Delta p^*[1 - 2v b_{hs}/\dot{V}^*] = 0.75\sqrt{\rho}\frac{\Omega_f v_f^{0.4}}{d^{1.4}}\frac{v^{1.6}}{(1-\alpha)^{1.6}}$$

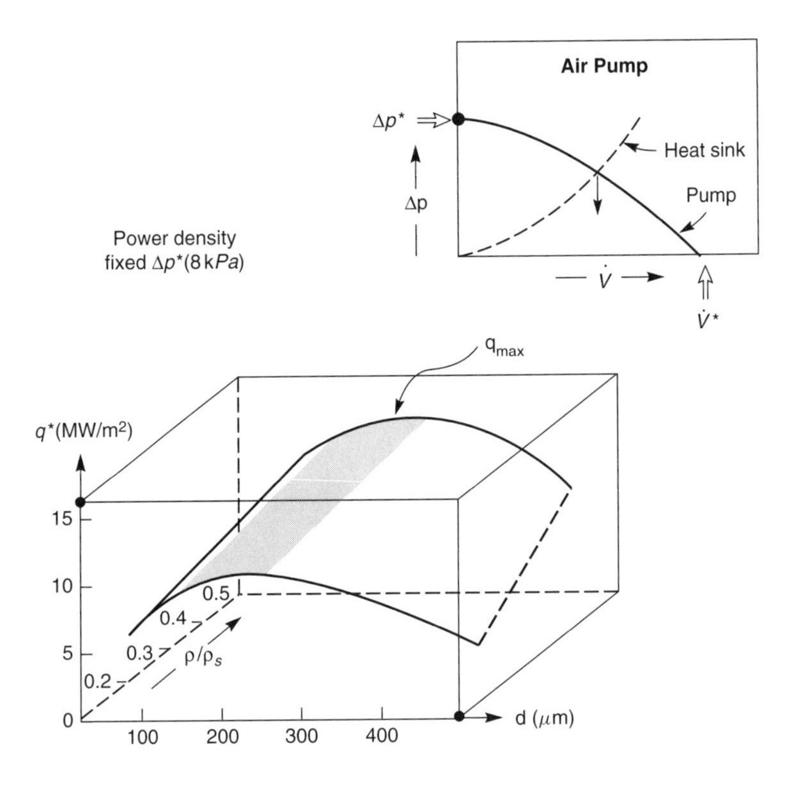

Figure 17.15 *Maximum on-chip thermal performance using mesocell metal heat sinks*

mance coefficients (Figure 17.15). These characteristics overlay with the pressure drop in the heat sink, also given in Table 17.1.

Equating the pressure drop with the operating characteristics results in an explicit flow rate for each heat sink (Figure 17.15): that is, for given cell size, relative density and thickness. With this defined flow rate, a specific heat flux, $\overline{q}$, can be accommodated by the design. Accordingly, for a prescribed fan, the heat sink exhibits a heat flux domain wherein the relative density and cell size are the variables. One such domain is indicated in Figure 17.15, calculated for $b_{hs} = 3\,\text{cm}$ and $b_{si} = 0.5\,\text{cm}$. Note that there is a ridge of high heat flux coincident with an optimum cell size. At a cell size smaller than this optimum, the pressure drop is excessive: conversely, at a larger cell size, the diminished heat transfer limits the performance. Along the ridge, there is a weak dependence of heat flux on relative density in the practical range ($\rho/\rho_s = 0.2\text{–}0.5$).

By selecting cellular materials that reside along the heat flux ridge, the requirements for the fan can be specified, resulting in a relationship between heat flux, back pressure and fluid flow rate. Some results for a representative density ($\rho/\rho_s = 0.3$) are plotted in Figure 17.16. The benefits of the cellular metal can only be utilized if the fan/blower assembly is capable of operating at back pressures of order 0.1 atm (10 kPa), while delivering flow rates about one l/s. Upon comparison with Figure 17.12, it is apparent that these heat fluxes substantially exceed those normally associated with forced air convection. A corollary of the heat flux is that there must be a temperature rise in the cooling air.

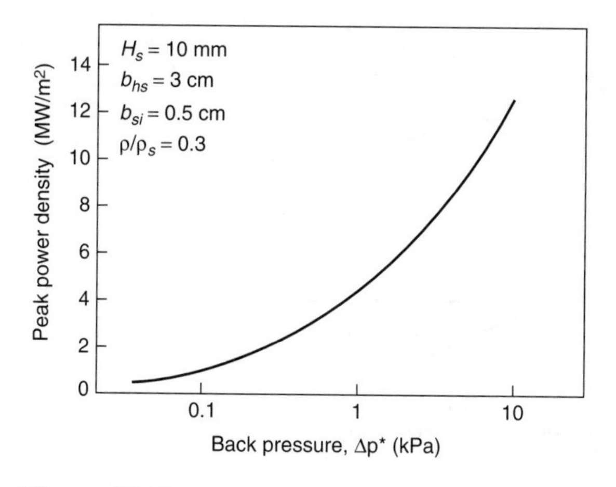

Figure 17.16 *Peak power density as a function of back pressure*

Device design issues

To take full advantage of the heat transfer capabilities of the cellular metal, the thermal design of the device must establish nearly isothermal conditions: that is, minimal thermal resistance between the electronics and the heat sink surface. This can be achieved by combining a high thermal conductivity aluminum nitride dielectric with a copper planar micro-heat-pipe, fully integrated with the heat sink by brazing (in order to exclude high thermal resistance at the interfaces*). Simulated temperatures for this scenario indicate that the required isothermality can be realized: albeit that the associated manufacturing requirements are stringent and yet to be demonstrated.

17.9 Applications under consideration

Skin-stiffened structures become more weight-efficient if the skin itself is a sandwich structure. This issue is discussed in some detail in Chapter 10, in

* Applications of polymer bonding and thermal greases would violate these requirements.

which optimized skin structures are compared, and illustrated by the case studies of Sections 17.1, 17.2 and 17.4. There are many more instances in which stiffness at low weight are sought. They include loudspeaker casings, display boards, overhead racks and folding tables in aircraft and in high-speed trains.

Furniture has to be light to be moveable; 'high-tech' fashion favors a metallic appearance. Metal foams with integral skins can be handled by many of the processes familiar to the furniture maker: cutting with bandsaws, joining with wood screws and adhesives, polishing to give attractive texture and surface finish. There appears to be potential for exploiting metal foams in furniture construction.

Cores for castings

Complex foam parts can replace sand cores used in foundry practice, to produce weight-saving cavities in casting. In this case the foam part will remain in the casting, saving labor and energy costs associated with the removal of the sand. In this way completely enclosed lightweight sections can be produced in castings which lead to significant improvements in mechanical, vibration and acoustic properties compared with the original hollow part.

Aesthetic applications

Metal foams appeal to industrial designers because of their surface texture, because they are novel (carrying associations of uniqueness) and because of their combination of light weight with bulk (giving reassurance of solidity in structures which are easily moved).

Chapter 18

Suppliers of metal foams

Certain information about foams – the availability, the exact price for a given quantity, delivery time and so forth – can only be obtained from the manufacturer, supplier or research organization. This section gives their names, addresses, phone, fax and e-mail numbers. Those with an asterisk (*) against their name make foams on a commercial scale. The others make foams – many with novel properties – but on a laboratory scale only, meaning that samples may be available for evaluation but large-scale supply is not, at present, possible.

Product name and contact information

*AEREX
Aerex Limited
Specialty Foams
CH-5643 Sins, Switerland
Tel: +0041 42 66 00 66
Fax: +0041 42 66 17 07

Duralcan (Dr L.D. Kenny)
Alcan International Ltd
Box 8400, Kingston
Ontario, K7L 5L9
Canada
Tel: +001 613 541 2400
Fax: +001 613 541 2134

*CYMAT (Dr Paul Ramsay)
Cymat Aluminum Corporation
1245 Aerowood Drive
Mississauga,
Ontario L4W 1B9
Canada
Tel: +001 905 602 1100 ext 17
Fax: +001 905 602 1250

e-mail: cymat@ican.net
Web: www.cymat.com

*ASTROMET
Astro Met Inc.
9974 Springfield, OH 45215
USA
Tel: +001 513 772 1242
Fax: +001 513 602 9080

*DUOCEL (Mr Bryan Leyda, Technical Director)
ERG Materials and Aerospace Corporation
900 Stanford Avenue
Oakland, CA 94608
USA
Tel: +001 510 658 9785
Fax: +001 510 658 7428
e-mail: sales@ergaerospace.com
Web: www.ergaerospace.com

*ALPORAS (Mr Maseo Itoh)
Shinko Wire Company Ltd
10-1 Nakahama-machi, Amagasaki-shi
660 Japan
Tel: +0081 6 411 1081
Fax: +0081 6 411 1056

*APORAS, Europe (Mr Karl Bula)
Innovation Services
CH-5200 Brugg
Herrenmatt 7F
Switzerland
Tel: +0041 56 422 5034
Fax: +0041 56 422 3635

*IFAM (Dr J. Banhart, Dr J. Baumeister)
Fraunhofer-Institute for Manufacturing and Advanced Materials
Wiener Strasse 12
D-28359 Bremen, Germany
Tel: +0049 421 2246 211
Fax: +0049 421 2246 300
e-mail: as@ifam.fhg.de

Fraunhofer-Delaware (Mr C.J. Yu)
Fraunhofer-Centre Delaware
501 Wyoming Road

Newark, DE 19716
USA
Tel: +1 302 369 6752
Fax: +1 302 369 6763
e-mail: yu@frc-de.fraunhofer.com

Fraunhofer-Dresden (Dr D.O. Anderson)
Fraunhofer-IFAM Dresden
Winterbergstraßae 28
D-01277 Dresden
Germany
Tel: +49 351 2537 319
Fax: +49 351 2537 399
e-mail: Anderson@epw.ifam.fhg.de

*SCHUNK (Dipl.-Ing. F. Baumgärtner)
Schunk Sintermetalltechnik GmbH
Postfach 10 09 51
D-35339 Gießen
Germany
Tel: +49 641 608 1420
Fax: +049 641 608 1734
e-mail: frank.baugaertner@schunk-group.com

*MEPURA (Professor P. Degischer, Director)
Mepura Metallpulvergesellschaft m.b.H.
Ranshofen
A-4813 Altmunster
Grossalmstrasse, 5, Austria
Tel: +0043 7612 88055-0
Fax: +0043 7612 88055-29
and
Metallpulvergesellschaft m.b.H.
Ranshofen
A-5282 Braunau-Ranshofen
Austria
Tel: +0043 7722 2216
Fax: +0043 7722 68154

*LKR (Professor P. Degischer, Director
Leichtmetall Kompetenzzentrum
Ranshofen
PO Box 26
A 5282 Ranshofen
Austria

Tel: +0043 7722 801-2125
Fax: +0043 7722 64393
e-mail: silberhumer@zdvaxf.arcs.ac.at

MEPURA-SLOVAKIA (Dr F. Simancik, Director)
Slovak Academy of Sciences
Department of Powder Metallurgy
PO Box 95
Racianska 75
Slovakia
Tel: (+42) 7 253000
Fax: (+42) 7 253301

*NEUMAN ALU FOAM
Neuman Alu Foam
A-3182 Marktl
Austria
Tel: (+43) 2762 500 670
Fax: (+43) 2762 500 679
e-mail: hoepler@neuman.at

GASAR
DML
Dnepropetrovsk Mettalurgical Institute
Ukraine

*HYDRO (Dr P. Asholt)
Hydro Aluminum, a.s.
R&D Materials Technology
PO Box 219
N 6601 Sunndalsøra
Norway
Tel: +0047 71 69 3000
Fax: +0047 71 69 3602

HOLLOW-SPHERE FOAM (Professor Joe Cochran)
Georgia Institute of Technology
Materials Science and Engineering
778 Atlantic Drive
Atlanta, GA 30332 USA
Tel: +001 404-894-6104
Fax: +001 404-894-9140
e-mail: joe.Cochran@mse.gatech.edu

JAM Corp. (Jon Priluc, President)
JAM Corp.

17 Jonspin Road
Wilmington, MA 01887-102
USA
Tel: +001 617 978/988 0050
Fax: +001 617 978/988 0080
e-mail: jpriluck@jamcorp.com

Molecular Geodesics Inc.
20 Hampden Street
Boston, MA 02199, USA
Tel: +001 617-427-0300
Fax: +001 617-427-1234
e-mail: resell@molec-geodesics.com

*INCO nickel foams (John B. Jones, Assistant Vice President)
INCO Limited
Research Laboratory
Sheridan Park
Mississauga
Ontario L5K 1Z9
Canada
Tel: +011 905 403 2465

*CFL nickel foams (Damien Michel)
Circuit Foil Luxembourg SA
PO Box 9
L-9501 Wiltz
G.D. of Luxembourg
Tel: +352 95-75-51-1
Fax: +352 95-75-51-249

ASHURST (Timothy Langan, Technical Director)
Ashurst Government Services
1450 S. Rolling Road
Balto, MD 21227, USA
Tel: +001 410-455-5521
Fax: +001 410-455-5500
e-mail: tlangan@ashurtech.com

Chapter 19

Web sites

An increasing number of Worldwide Web sites carry useful information about metal foams. Section 19.1 catalogues those of research groups and institutions engaged in research on metal foams. Section 19.2 lists those of suppliers. Section 19.3 contains other relevant sites. All the sites listed here were tested and were active on 1 August 1999. The Web is in continuous flux, so that the list must be seen as a temporal snapshot of sites, not a permanent record.

19.1 Web sites of academic and research institutions

http://das-
www.harvard.edu/users/faculty/Anthony_Evans/Ultralight_Conference/
UltralightStructures.html
Division of Applied Sciences, Harvard, Ultralight Structures programme

http://das-
www.harvard.edu/users/faculty/Anthony_Evans/Ultralight_Conference/
MetalFoams.html
http://das-
www.harvard.edu/users/faculty/Evans/technical_articles/Article3/Article3.html
Details of papers on metal foams from the Harvard-based MURI project

www.ipm.virginia.edu
University of Virginia, Materials Science and Engineering

http://www.ifam.fhg.de/fhg/ifam/e_ifoam.html
http://www.ifam.fhg.de/fhg/ifam/e_paperfoam.html
IFAM (Fraunhofer Institute, Bremen) web site for metal foam research

http://www.mse.gatech.edu/sphere.html
Information on titanium and steel hollow-sphere foams developed at Georgia Tech.

http://www.msm.cam.ac.uk/mmc/project.html#vlado
Metals foam research at Cambridge, UK

http://hotmetals.ms.nwu.edu/Dunand/metal_foams.html
Northwestern University, Illinois, USA, research on foamed titanium

http://ilfb.tuwien.ac.at/ilfb/abstr/GRa99.html
Research at the Technical University of Vienna on energy management with metal foams

http://www.ipm.virginia.edu/research/PM/Pubs/pm_pubs.htm
Research at the University of Virginia on metal foams

http://www.mpif.org/aeros.html
Table of contents of a book, one 8-page chapter of which concerns 'Weight saving by aluminum metal foams: production, properties and applications in automotive'

http://mgravity.itsc.uah.edu/microgravity/micrex/ComGases.html
Links to abstracts of papers written by J. Pratten in the mid-1970s on metal foams at the Marshall Space Flight Center for Microgravity Research

http://aries.ucsd.edu/~flynn/EMEET/TILLACK/ultamet.html
Ultramet foams

http://www.brenner.fit.edu/metfoams.htm
Metal foams used for enhanced heat transfer in chemical engineering.

http://www.zarm.uni-bremen.de/2forschung/ferro/ferrofluid.html
Foams made under low-gravity conditions

http://silver.neep.wisc.edu/~lakes/PoissonFas.html
Information on negative Poisson ratio foams

19.2 Web sites of commercial suppliers

http://ergaerospace.com/
ERG Materials and Aerospace is a manufacturer of open-cell aluminum and vitreous carbon foam materials for the aerospace, transport and semiconductor markets.

http://ergaerospace.com/lit.html.
ERG Composite Literature and Reports: *Duocell Aluminum Foam* and *Duocell Physical Properties*. See Chapter 17 for contact information.

http://www.cymat.com
CYMAT, supplier of aluminum-based closed-cell foams. See Chapter 17 for contact information.

http://www.hydro.com
Norsk Hydro, supplier of aluminum-based open-cell foams. See Chapter 17 for contact information.

http://www.astromet.com/200.htm
Astromet are producers of metal foams and other advanced materials.

http://www.seac.nl/english/recemat/index.html
SEAC RECEMAT open-cell nickel-based foam producers.

http://www.bellcomb.com/examples.htm
Producers of metal honeycomb materials.

http://www.molec-geoedesics.co..htm
Producers of lattice-structured preforms.

19.3 Other web sites of interest

http://www.karmann.de/d/d_2/index_d_2.htm
Karmann GmbH web site describing innovative use of metal foams in concept vehicles

http://www.ultramet.com/7.htm
Details of methods of making lightweight foams

http://www.dedienne.dedienne.com/fiches/materiauxuk.htm
Advanced materials listing

http://www.shelleys.demon.co.uk/fea-sep.htm
Materials magazine article highlighting metal foam manufacturing methods

http://www.designinsite.dk/htmsider/m0153.htm
Design InSite metal foams information page

http://www.ifam.fhg.de/fhg/ifam/meteor.html
Lightweight metal foam research project funded by the European Community

Appendix: Catalogue of material indices

Material indices help to identify the applications in which a material might excel and those for generic components are assembled here. Their derivation and use is illustrated in Chapters 5, 11 and 12.

(a) Stiffness-limited design at minimum mass (cost, energy[a])

Function and constraints[a]	Maximize[b]
SHAFT (loaded in torsion)	
Stiffness, length, shape specified, section area free	$G^{1/2}/\rho$
Stiffness, length, outer radius specified; wall thickness free	G/ρ
Stiffness, length, wall thickness specified, outer radius free	$G^{1/3}/\rho$
BEAM (loaded in bending)	
Stiffness, length, shape specified; section area free	$E^{1/2}/\rho$
Stiffness, length, height specified; width free	E/ρ
Stiffness, length, width specified; height free	$E^{1/3}/\rho$
COLUMN (compression strut, failure by elastic buckling) buckling load, length, shape specified; section area free	$E^{1/2}/\rho$

[a]To minimize cost, use the above criteria for minimum weight, replacing density ρ by $C_m\rho$, where C_m is the material cost per kg. To minimize energy content, use the above criteria for minimum weight replacing density ρ by $q\rho$ where q is the energy content per kg.
[b]E = Young's modulus; G = shear modulus; ρ = density.

(b) Stiffness-limited design at minimum mass (cost, energy[a])

Function and constraints[a]	Maximize[b]
PANEL (flat plate, loaded in bending) stiffness, length, width specified, thickness free	$E^{1/3}/\rho$
PLATE (flat plate, compressed in-plane, buckling failure) collapse load, length and width specified, thickness free	$E^{1/3}/\rho$

(b) (*continued*)

Function and constraints[a]	Maximize[b]
CYLINDER WITH INTERNAL PRESSURE	
elastic distortion, pressure and radius specified; wall thickness free	E/ρ
SPHERICAL SHELL WITH INTERNAL PRESSURE	
elastic distortion, pressure and radius specified, wall thickness free	$E/(1-\nu)\rho$

[a]To minimize cost, use the above criteria for minimum weight, replacing density ρ by $C_m\rho$, where C_m is the material cost per kg. To minimize energy content, use the above criteria for minimum weight replacing density ρ by $q\rho$ where q is the energy content per kg.
[b]E = Young's modulus; G = shear modulus; ρ = density.

(c) Strength-limited design at minimum mass (cost, energy[a])

Function and constraints[a,c]	Maximize[b]
SHAFT (loaded in torsion)	
Load, length, shape specified, section area free	$\sigma_f^{2/3}/\rho$
Load, length, outer radius specified; wall thickness free	σ_f/ρ
Load, length, wall thickness specified, outer radius free	$\sigma_f^{1/2}/\rho$
BEAM (loaded in bending)	
Load, length, shape specified; section area free	$\sigma_f^{2/3}/\rho$
Load, length, height specified; width free	σ_f/ρ
Load, length, width specified; height free	$\sigma_f^{1/2}/\rho$
COLUMN (compression strut)	
Load, length, shape specified; section area free	σ_f/ρ
PANEL (flat plate, loaded in bending)	
Stiffness, length, width specified, thickness free	$\sigma_f^{1/2}/\rho$

[a]To minimize cost, use the above criteria for minimum weight, replacing density ρ by $C_m\rho$, where C_m is the material cost per kg. To minimize energy content, use the above criteria for minimum weight replacing density ρ by $q\rho$ where q is the energy content per kg.
[b]σ_f = failure strength (the yield strength for metals and ductile polymers, the tensile strength for ceramics, glasses and brittle polymers); ρ = density.
[c]For design for infinite fatigue life, replace σ_f by the endurance limit σ_e.

(d) Strength-limited design at minimum mass (cost, energy[a])

Function and constraints[a,c]	Maximize[b]
PLATE (flat plate, compressed in-plane, buckling failure) Collapse load, and width specified, thickness free	$\sigma_f^{1/2}/\rho$
CYLINDER WITH INTERNAL PRESSURE Elastic distortion, pressure and radius specified; wall thickness free	σ_f/ρ
SPHERICAL SHELL WITH INTERNAL PRESSURE Elastic distortion, pressure and radius specified, wall thickness free	σ_f/ρ
FLYWHEELS, ROTATING DISKS Maximum energy storage per unit volume; given velocity Maximum energy storage per unit mass; no failure	 ρ σ_f/ρ

[a]To minimize cost, use the above criteria for minimum weight, replacing density ρ by $C_m\rho$, where C_m is the material cost per kg. To minimize energy content, use the above criteria for minimum weight replacing density ρ by $q\rho$ where q is the energy content per kg.
[b]σ_f = failure strength (the yield strength for metals and ductile polymers, the tensile strength for ceramics, glasses and brittle polymers); ρ = density.
[c]For design for infinite fatigue life, replace σ_f by the endurance limit σ_e.

(e) Strength-limited design: springs, hinges etc for maximum performance[a]

Function and constraints[a,c]	Maximize[b]
ELASTIC HINGES Radius of bend to be minimized (max. flexibility without failure)	σ_f/E
COMPRESSION SEALS AND GASKETS Maximum conformability; limit on contact pressure	$\sigma_f^{3/2}/E$ and $1/E$
ROTATING DRUMS AND CENTRIFUGES Maximum angular velocity; radius fixed; wall thickness free	σ_f/ρ

[a]To minimize cost, use the above criteria for minimum weight, replacing density ρ by $C_m\rho$, where C_m is the material cost per kg. To minimize energy content, use the above criteria for minimum weight replacing density ρ by $q\rho$ where q is the energy content per kg.
[b]σ_f = failure (the yield strength for metals and ductile polymers, the tensile strength for ceramics, glasses and brittle polymers); H = hardness; ρ = density.
[c]For design for infinite fatigue life, replace σ_f by the endurance limit σ_e.

(f) Vibration-limited design

Function and constraints[a,c]	Maximize[b]
TIES, COLUMNS	
Maximum longitudinal vibration frequencies	E/ρ
BEAMS	
Maximum flexural vibration frequencies	$E^{1/2}/\rho$
PANELS	
Maximum flexural vibration frequencies	$E^{1/3}/\rho$
TIES, COLUMNS, BEAMS, PANELS	
Minimum longitudinal excitation from external drivers, ties	$\eta E/\rho$
Minimum flexural excitation from external drivers, beams	$\eta E^{1/2}/\rho$
Minimum flexural excitation from external drives, panels	$\eta E^{1/3}/\rho$

[a]To minimize cost, use the above criteria for minimum weight, replacing density ρ by $C_m\rho$, where C_m is the material cost per kg. To minimize energy content, use the above criteria for minimum weight replacing density ρ by $q\rho$ where q is the energy content per kg.
[b]σ_f = failure (the yield strength for metals and ductile polymers, the tensile strength for ceramics, glasses and brittle polymers); η = damping coefficient; ρ = density.
[c]For design for infinite fatigue life, replace σ_f by the endurance limit σ_e.

(g) Thermal and thermo-mechanical Design

Function and constraints	Maximize[a]
THERMAL INSULATION MATERIALS	
Minimum heat flux at steady state; thickness specified	$1/\lambda$
Minimum temp rise in specified time; thickness specified	$1/a = \rho C_p/\lambda$
Minimize total energy consumed in thermal cycle (kilns, etc.)	$\sqrt{a}/\lambda = \sqrt{1/\lambda\rho C_p}$
THERMAL STORAGE MATERIALS	
Maximum energy stored/unit material cost (storage heaters)	C_p/C_m
Maximize energy stored for given temperature rise and time	$\lambda/\sqrt{a} = \sqrt{\lambda\rho C_p}$
PRECISION DEVICES	
Minimize thermal distortion for given heat flux	λ/α
THERMAL SHOCK RESISTANCE	
Maximum change in surface temperature; no failure	$\sigma_f/E\alpha$

(g) (*continued*)

Function and constraints	Maximize[a]
HEAT SINKS	
Maximum heat flux per unit volume; expansion limited	$\lambda/\Delta\alpha$
Maximum heat flux per unit mass; expansion limited	$\lambda/\rho\Delta\alpha$
HEAT EXCHANGERS (pressure-limited)	
Maximum heat flux per unit area; no failure under Δp	$\lambda\sigma_f$
Maximum heat flux per unit mass; no failure under Δp	$\lambda\sigma_f/\rho$

[a] λ = thermal conductivity; a = thermal diffusivity; C_p = specific heat capacity; C_m = material cost / kg; T_{max} = maximum service temperature; α = thermal expansion coeff; E = Young's modulus; ρ = density; σ_f = failure strength (the yield strength for metals and ductile polymers, the tensile strength for ceramics, glass and brittle polymers).

Index

(Company and trade names are capitalized)